DAZZLE, DISRUPTION & CONCEALMENT

DAZZLE, DISRUPTION & CONCEALMENT

THE SCIENCE, PSYCHOLOGY & ART OF SHIP CAMOUFLAGE

DAVID L. WILLIAMS

The
History
Press

The T3-S2-A1 auxiliary oiler USS *Mississinewa* in 1994, in Measure 32 Design 3AO camouflage. (US Naval History and Heritage Command, NH97279)

First published 2022

The History Press
97 St George's Place, Cheltenham
Gloucestershire, GL50 3QB
www.thehistorypress.co.uk

British Library Cataloguing in Publication Data.
A catalogue record for this book is available from the British Library.

ISBN 978 0 7509 9681 5

Typesetting and origination by The History Press
Printed in Turkey by Imak

Trees for Life

CONTENTS

Art is never twice the same; Science has no variables. We pay tribute to the artist pioneers here and abroad, the men who hopefully and daringly led, while Science slowly wrought for truth and towards safety. They were fantastic sometimes these men, fertile in amazing designs, prolific in output. But they made a bridge to better things. Ship-owners and underwriters henceforth know how to get the safety that lies in Camouflage.

Artists have scope for talent in the varied combinations of deceptive lines, curves, stripes and patterns. The manufacturers can produce the right paints and pigments. The mystery is lifted for the master-painter.

Lindon W. Bates, Chairman of the Engineering Committee,
Submarine Defense Association

INTRODUCTION

Essentially, prior to the First World War, the concept of camouflage or protective colouration, other than in the natural world, had only been implemented on rare occasions in the distant historical past. It was the changing character of warfare and weaponry in recent times that demanded the adoption of a comprehensive means of visual secretion for fighting forces and all forms of military equipment as well as strategic structures. This book is concerned with painted ship camouflage in all its forms – concealment, disruption and deception – as it emerged and evolved from the final years of the nineteenth century.

Static objects can generally be hidden using traditional camouflage methods, by matching them to their background, and it is only through movement that they run the risk of detection. This applies to most military vehicles and land forces personnel. But it was never the case for ships on the sea which, according to their purpose, are obliged to move constantly in a highly revealing environment.

The term 'camouflage' is widely recognised and most people intuitively understand what it refers to, but in reality, where ships are concerned, it can be misleading, and protective colouration in the maritime world can mean a great deal more. The term 'camouflage' originates from a French verb, '*camoufler*', meaning to hide, deceive or disguise. As such, it largely refers to a single stratagem for the protective colouration of ships – background blending – yet, it has become an all-embracing term, adopted internationally as the name for any system conceived for that purpose and, for that reason, it is used throughout this book to refer to all forms of painted protection.

Drawing on numerous official reports, original specifications and contemporary accounts, the primary aim of this book is to present the story of the quest for effective camouflage for ships at sea. As part of that, quite apart from presenting a broad spectrum of photographs to illustrate various forms of painted protection measures, it addresses certain central themes.

First, it focuses on the evolution of ship camouflage schemes, with emphasis on the scientific and mathematical principles involved in addition to the artistic dimensions. The detailed specifications (the designs, patterns and paint colours) of particular concealment or disruption schemes are not covered here, the central theme being their conception from embryonic origins, influenced by concealment in nature, to the subsequent attempts to apply behavioural and scientific erudition to achieve certain objectives. It also focuses on the behind-the-scenes processes of trial and experimentation of concepts that led ultimately to practical implementation. There is a good reason for adopting this approach. In the *Oxford Handbook of Perceptual Organization*, in a thesis by Daniel Osorio and Innes C. Cuthill on the subject of camouflage in the animal kingdom, the following statement is made:

Recent research on camouflage has been stimulated by the realization that direct evidence for how particular types of camouflage exploit perceptual mechanisms was sparser than textbooks might suggest. Also, such evidence as did exist had been evaluated via human perception … not 'the exploitation of the functionality of the physical sciences'.

This observation, recognising the need for a balance between the workings of visual perception and scientific behaviour, is equally applicable to the endeavours to develop camouflage for ships at sea.

Secondly, this book identifies all the key individuals and their contributions to the better scientific understanding of the factors involved in painted protection at sea. It looks at the tools, apparatus and techniques they contrived, for both design and test, as well as the terminology, definitions and appraisal criteria that were devised, all to permit the development of effective strategies, as well as the reliable, accurate and comparative evaluation of the various schemes proposed. The organisational framework in which camoufleurs (camouflage designers and developers) functioned is also explained, both the establishments that were created and their hierarchical relationships as well as the inevitable conflicts that arose between certain of them.

In the chapters that follow, the rationale behind each of the broad camouflage strategies proposed during the First World War is explored, along with its development through to its introduction. This is followed, as a third core element, by a review of the available evidence from contemporary inquiries and studies, in an attempt to reach dependable conclusions as to the effectiveness or otherwise of certain of the measures that were adopted.

Marine camouflage is the most complex subject and presents an incredible challenge to achieve. To reinforce what has already been stated, a ship at sea cannot be easily concealed, unlike a tank or a howitzer gun on land. The ship naturally stands out against an ever-changing background of sea and sky, the variations of which are innumerable: the level of light and visibility, the angle and direction of illumination, the state of the weather and the ceaselessly varying aspect of the sea, all influence the canvas against which ships are viewed. Further complications arise from the ship's inherent structure of deck overhangs, masts, cowls, lifeboats in davits and so on, besides smoke emissions from the funnels and reflections off the hull plates – all difficult, if not impossible to hide.

The need for a defensive mechanism for ships in the form of painted colour or pattern was first given serious deliberation in the USA in the closing years of the nineteenth century. At that time, the colours that naval ships were painted in was already in a state of flux, driven by developments in naval weaponry. For as long as ships were in close proximity when they engaged in battle, the colours in which they were painted were of little consequence.

Up to the end of the Victorian era, Royal Navy ships had black hulls, white upperworks and buff funnels. The warships of the USA were known as the 'Great White Fleet' for obvious reasons, and similar brightly coloured paint schemes were followed by many other navies. But with the increasing range of naval artillery, visibility at a distance became an issue and, from that time, an overall coat of mid-tone grey in various shades and tones was progressively adopted in recognition of the fact that it offered a degree of concealment, besides being a utility service livery that could be conveniently maintained.

However, it was recognised early on that this approach could only provide visual protection in a single set of circumstances, when the ship's colour corresponded to the colour of the background against which it was viewed. Short of having chameleon-like properties, able to adapt to changing levels of light, haze and cloud cover, ships would, by definition, be potentially more conspicuous at all other times.

Nonetheless, the situation remained largely unchanged at the outbreak of the First World War. However, the war was to demand much more of this form of protection, both for mercantile and naval vessels. In the face of a deadly weapon combination – the submarine and self-propelled torpedo – a grey coating alone would prove to be inadequate.

In both world wars, the UK was in a unique predicament. With its expansive empire and fundamental dependence on the supply of food, fuel and other essential commodities from overseas, plus the manpower required to support its war effort, it was essential to keep the ocean supply lines open. The protection of its vulnerable merchantmen, on which this depended, was to prove an immense challenge for, in both periods of hostilities, enemy submarines wreaked havoc on the ocean highways, especially during the first global conflict.

Initially, the UK placed much trust in Germany's respect of and adherence to the principles of the cruiser or prize rules, which had evolved since the seventeenth century, ultimately taking the form of an international agreement from 1909 (the London Declaration), which defined the maritime law applicable during times of conflict. It was soon discovered that this was not to be the case, though, and German unrestricted submarine warfare rapidly took a heavy toll on British and Allied shipping.

The fact was that none of the signatories had ever ratified their adoption of this declaration, which, anyway, was only applicable to the conduct of surface warships. In practice, the 'cruiser rules' were only drawn up to protect ships' crews, not the ships themselves, which, once forewarned and abandoned, could be legally sunk or confiscated with their valuable cargoes. Besides, the scope of the declaration was indefinite where submarines were concerned, and according to their modus operandi they would have lost their military value if the intention to sink was communicated to the target in advance of launching an attack. If anything, the First World War revealed that combat was no longer a business conducted according to some 'gentlemanly' rules of behaviour but had descended to an uncivilised level with little regard for who or what fell victim, be it civilian or military.

The grave situation that ensued at sea in 1915 was initially allayed when Germany was compelled to restrain its underwater offensive in the face of retaliatory consequences threatened by the USA. But, to avoid the risk of its own defeat, Germany abrogated those undertakings and recommenced unrestricted submarine warfare. Attacks and sinkings worsened alarmingly from the spring of 1917, with record numbers of merchant ships sunk every week and, starved of food and supplies, the UK came disastrously close to collapse. Something had to be done. The critical situation facing the nation dictated a need to urgently adopt radical solutions.

As part of its response, the UK Government reintroduced the convoy system as one essential action to tackle the acute predicament. Another was the introduction of a system of protective colouration of ships, which was completely at odds with anything that had been advocated previously. If concealment within the ocean and sky background to reduce visual prominence, as had already been attempted, could not always be achieved, emphasis instead would be placed on interfering with the judgement of target distance, speed and bearing. Known as 'dazzle', this was a system of disruptive painting which was intentionally designed *not* to conceal a ship, in the belief that such an objective was impossible, but to so confuse the commander of an attacking submarine that he would aim his torpedo incorrectly and miss the target. It was controversial, the complete antithesis of concealment and, as it turned out, the catalyst for a rivalry that developed between the schools of marine camouflage strategy.

Although the efforts to formulate methods of effective camouflage in the maritime environment did not originate in the UK, from early 1917 it was the British authorities who were to lead the way, no doubt because of the country's critical dependence on sea trade, exhibiting an uncharacteristic degree of flexibility and accommodation, considering the conservative tendencies for which the Admiralty was known. But, once the USA entered the war, the focal point of marine camouflage development began to shift, a realignment that continued through the inter-war period and into the Second World War.

So, were the pioneering naval camouflage schemes introduced in the First World War genuinely effective and, if so, were they retained? Continued and widespread use of ship camouflage practices during the Second World War would suggest that they were – but were the schemes adopted in this second conflict the same? Or were they some form of derivative or the product of completely fresh thinking?

It is hoped that the reader's appreciation of the complexities of maritime camouflage will be enhanced through the perspective this book brings to the subject, seeking to understand the scientific principles and visual mechanisms that naturalists, artists and physicists sought to utilise in its investigation and development.

David L. Williams, January 2022

SCIENTIFIC PRINCIPLES AND TECHNICAL TERMS PERTAINING TO CAMOUFLAGE

It is not the intention for this book to be overly technical. However, it is thought that the reader would benefit from simplified explanations of certain scientific principles and other technical terms that are directly relevant to camouflage conception and development in order to appreciate the rationales that lay behind different maritime camouflage strategies.

Three core physical processes are central to an appreciation of camouflage practices – **light**, **colour** and **vision**. Together, the science of light and the nature of colour, along with the physiology of vision, may be regarded as the essential components exploited by camoufleurs in their attempts to render ships less visible or make them less easy to attack successfully. These factors are equally applicable to all forms of military camouflage but they have greater pertinence where maritime camouflage is concerned because of the peculiar obstacles to concealment or deception in the marine environment.

LIGHT AND COLOUR

Essentially, without light there would be no vision or colour. **Visible light** is a form of vibrating or oscillating energy that radiates in waves from a source such as the sun, travelling through space until it impinges on a surface. As a segment of a much larger electromagnetic spectrum of energy, it comprises a **spectrum** of the familiar colours of the rainbow, known typically, reading from the longest wavelength to the shortest, as red, orange, yellow, green, blue, indigo and violet.

Light energy has essential properties which are important to camouflage development. It travels extremely fast and in straight lines. The latter property means it can be blocked, leaving bright (illuminated) surfaces and shaded (unilluminated) surfaces. As a result, the appearance of ships at sea can change significantly and perceptively as they steer a course or pitch and roll in the ocean waves, changing their attitude in relation to the source of illumination and the position of the observer.

Camoufleurs are interested in how light behaves when it impinges on surfaces and how mixtures of different wavelengths of light energy behave when juxtaposed in combination.

VISION

Vision is one of our five senses by which we relate to the world around us. Isaac Newton stated that vision is a combination of physical phenomena and perceptual phenomena. Each is described here in the context of its influence on camouflage development. Physical phenomena may be divided as having external and internal dimensions – in the former case, how objects react to light, and in the latter, how the eye works physiologically to process the visual data.

External Physical Phenomena

When light impinges on a surface, some wavelengths are absorbed (**absorption**), while others are reflected (**reflection**). The individual physical properties of each object determine how this will occur and in what ratios. Put another way, objects assume a colour which is a mixture of those spectral wavelengths that are present in the light from an illuminating source which they do not absorb.

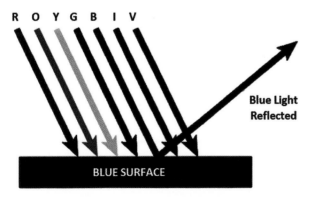

A simplified diagram showing the process whereby objects assume a colour, in this case, blue. All the wavelengths of incident white light are absorbed other than blue, which is reflected. Total reflection of all wavelengths would result in a white surface; total absorption of all wavelengths would result in a black surface. (Author)

It follows that if the composition of the wavelengths of the light incident upon the object do not permit it to react in accordance with its characteristic attributes, it will appear different to what may be normally expected.

Reflected light has other important properties. In **specular reflection**, light is reflected as if from a mirror or shiny surface, in a straight line at an angle equal and opposite to the angle of incidence. In **diffuse reflection**, the light is scattered in many directions and will be seen as a glow rather than a mirror image. Matt surface coatings, essential for camouflage, behave in this way.

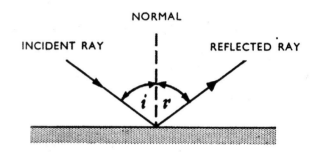

Light ray reflected by a plane mirror

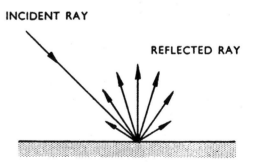

Light ray reflected by a matt surface

The difference between specular (top) and diffuse (bottom) reflection. It is only by diffuse, reflected light that detail and texture can be perceived. This dictated a requirement for matt- or dull-finish paints to be applied for ship camouflage in order to reduce specular reflections to the minimum. (Author)

Internal Physical Phenomena

Equally important is how the human eye reacts to visual stimuli. Without going into the complexities of the workings of the human eye or how the experience of vision works, it is sufficient to say here that the eye's retina is made up of two types of **photoreceptors**, known as **rods** and **cones**, which respond to light stimuli.

The cones are characterised by having lower light sensitivity but support high **acuity** (sharpness). This is known as **photopic** or **daylight vision**.

Conversely, the rods have greater light sensitivity but poor acuity. Called **scotopic vision**, this is how we see when there is less light, such as at twilight when the eye undergoes **dark adaptation** as a means of assisting vision at lower light levels.

How these different photoreceptors work has a bearing on how visual information is interpreted and the exploitation of this behaviour was central to certain of the attempts to create effective ship camouflage.

A further, important quality of the rods and cones of the retina relates to the sensation of colour as postulated by the trichromatic (red/green/blue) theory of colour vision in which the various rod and cone photoreceptors exhibit different responses in their spectral sensitivity, their response to the different wavelengths of the spectrum. One of these differences is manifest in a curious perceptual phenomenon that occurs as the light level varies and vision switches from photopic to scotopic and vice versa. Known as the **Purkinje Shift**, it is a deviation of visual perception that offered valuable potential when applied to the selection of camouflage paint colours.

In simple terms, in photopic (daylight) vision, objects whose colour attributes fall in the blue-green wavelengths appear to be relatively lighter than those that have yellow-green attributes. The phenomenon reverses when vision is scotopic (low-level light), resulting in blue-green colours appearing to be darker than those that are yellow-green. In practical terms, the Purkinje Shift could be exploited in camouflage applications because green paints could inherently appear either darker or lighter according to the level of light.

Visual acuity, the resolving power of the eye, also has a big influence on camouflage design. As has already been stated, the cone photoreceptors assist sharpness of vision and **resolution** of detail. This ability determines how camouflage systems will be perceived but, apart from the capabilities of the eye, resolution of the detail in an object is also governed by the **granularity** of that detail and the distance from which it is observed.

For each patterned design, there is a point at which the eye (or the assisted eye, through binoculars) is unable to resolve the pattern detail and it will then be seen as a uniform tone. This is known as the **blending distance**. The inability to resolve detail at a distance can be simply demonstrated by looking very closely at or magnifying a photograph printed in a book or newspaper. It will be seen to be made up of only pure black dots of varying size on the white paper. By holding the image at arm's length only grey tones will be observed because the dot pattern has blended at a distance beyond the eye's ability to resolve the detail. In the design of camouflage patterns, camoufleurs are interested in what the observer will see either side of the blending distance.

When considering the pattern detail of an object, it should be noted that the pattern comprises two elements if the object, in this instance a ship, has been painted in a camouflage design. There is the **structural pattern**, inherent to the ship – the details of light and shade arising from its deck overhangs, fixtures and fittings – and the **painted pattern** of the design, applied to the ship in accordance with the prescribed camouflage measure. Together, the structural pattern and the painted pattern combine to become the **apparent pattern**.

Perceptual Phenomena

The complexities of the physical aspects of vision are complicated further by **visual perception** – what we believe we see, irrespective of any physical shortcomings such as colour blindness or lens defects, rather than what we actually see. There are various mechanisms which help us make sense of what we see. Seeing through two eyes enables us to experience the phenomenon of **stereopsis**, or binocular vision, by which we can see objects as three-dimensional, allowing us to determine their depth and dimensions, judge their distance away from us and their relationship in space to other objects.

An illusion of depth can be simulated through the effect of **linear** or **vanishing point perspective**, whereby a two-dimensional image, as in a photograph or painting, will be seen to be three-dimensional, with some elements apparently nearer while others are further away. Camoufleurs endeavoured to deceive observers by replicating this effect on the hulls of ships, employing the principles of both **plane** and **solid geometry**, to deliberately distort true perspective and, as a result, interfere with the estimation of size, shape, distance and inclination.

Another important psychological aspect of vision which can have a bearing on how we interpret what we see and which ship camouflage designers sought to take advantage of is explained by what is known as **Gestalt Theory**. It would be inappropriate here to try to explain this complex theory in any depth. Suffice to say that one of the principal tenets of Gestalt Theory, manifest as the principle of organisation, states, 'The attributes of the whole are not deducible from analysis of the parts in isolation.'

What this means, in practical terms, is that an entire object cannot necessarily be deduced when only parts of it can be seen separately. Organisation is a psychological dimension of the visual process, which helps make objects comprehensible. For recognition of an object to occur, it is required to be segregated from its surroundings by stimulus properties that group its constituent parts together as a coherent whole or single perceptual unit.

Fragmentation of this organisational continuity was sought in marine camouflage by deliberate exploitation of intense colour contrast to cause the breakdown of the cohesion or grouping of the constituent parts of a ship. An illusion of selective disappearance could thus be achieved through the apparent elimination of parts of the discernible structure by tonally associating those parts with the background, in effect, 'hiding' them from view.

In the context of camouflage practices for ships, all these aspects of the laws of physics and visual perception were open to manipulation. The old adage is 'seeing is believing', which is exactly what the camouflage designer or technician wanted the observer to do when he employed visual trickery to convey misinformation.

WEATHER

Apart from the physics of light, colour and vision, the impact of weather conditions should be mentioned, insofar as they affect camouflage observation. Specifically, it is the effect of atmospheric humidity that is of primary interest as this has a direct bearing on the conspicuity of a vessel at sea. The term of importance here is **haze**, which is the optical effect of mist, the optical scattering of material present in the atmosphere, which is more prevalent in certain sea areas than in others.

A BRIEF NOTE ON THE INTERPRETATION OF PHOTOGRAPHS

At the outbreak of the Second World War, colour photography was still in its infancy and few images were recorded of ships by that medium to show the actual colours of camouflage livery. The vast majority of the photographs from that conflict and all the photographs taken in the First World War were in monochrome (black and white). However, two types of monochrome film were in general use during the First World War, which recorded colours in different ways. They were **orthochromatic** (blue and green sensitive) and **panchromatic** (blue, green and red sensitive).

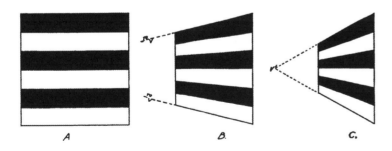

The illusion of stereopsis contrived with lines of perspective or patterned texture gradients, here suggesting that a 2D illustration has depth or three dimensions. The sense of depth increases the more the lines are angled, as shown by comparing figures B and C. (Alon Bement)

In the same way as light which lacks certain wavelengths can affect the colour appearance of an object, so too the lack of sensitivity to certain wavelengths of light by a photographic film emulsion will also record the colour appearance of an object differently. With orthochromatic film, areas of the object in colours at the red end of the spectrum will not record on a negative and, as a result, will appear darker than they are in a print positive subsequently made from that negative.

Further complications affecting the reproduction of tones in black and white photographs arise from deviations from correct negative exposure (under or over) and from the selection of photographic paper contrast (either soft or hard) when making prints. As the relative tones of grey of a camouflage design may be distorted through the film behaviour or exposure or printing contrast variations, the viewer should exercise caution when judging the apparent appearance of camouflage systems as seen in photographs.

While the photographs reproduced in the book may prove useful to those readers who are model makers, their presence here is not intended for that purpose.

1

HARD TO SEE, HARD TO HIT – LOW VISIBILITY 1

The aim of camouflage is generally understood to be the concealment of an object against or within its surroundings, so that it cannot be seen. If complete invisibility cannot be achieved, the next best thing, logically, is to seek reduced or low visibility. This was the strategy that dominated the thinking of those who sought a means of painted protection for ships both before and during the early years of the First World War.

Three significant events that occurred between 1896 and 1902 signalled the beginning of true camouflage development for ships, then referred to as 'marine protective colouration'. These were, first, the publication in April 1896 of a paper by Abbott Handerson Thayer, titled 'The Law Which Underlies Protective Coloration', now known as 'Thayer's Law', being an explanation of the system by which the colouring of living creatures has evolved to provide protective concealment, a system which Thayer felt could be applied beyond the natural world. It was followed up two years later, at the time of the Spanish–American War, when Thayer, on the invitation of the US Navy Department, demonstrated the principles of the scheme he had devised for concealing ships, an exercise which, in the

event, proved to be inconclusive. This culminated in the registration by Thayer, jointly with Gerome Brush, of US Patent No. 715,013, 'Process of Treating the Outsides of Ships Etc., For Making Them Less Visible', dated 2 December 1902.

Together, these events not only represented the apotheosis of the studies and experiments into defensive colouration in nature conducted by Abbott Handerson Thayer since the early 1890s, but also the first suggestions that the principles of the behaviour he had observed – counter-shading and background blending (sometimes referred to as 'protective resemblance') – could be applied to the structures of ships. Thayer, a naturalist and artist from New Hampshire, USA, has since been referred to reverentially as the 'father of camouflage'.

Seven years later, Gerald H. Thayer, his son, published a book entitled *Concealing Coloration in the Animal Kingdom – An Exposition of the Laws of Disguise Through Color and Pattern: Being a Summary of Abbott H. Thayer's Disclosures*, a text which reinforced the claims that had been made in the patent and added to Thayer's reputation as an expert in this field. Essentially, Thayer was advocating the adoption of certain specific practices for ship camouflage which emanated from his prolonged observation of animal behaviour, the focus being entirely on reducing visibility.

Foremost was the concept of Counter-Shading, known as Thayer's Law. Predicated, over-ambitiously, as a universally applicable phenomenon in nature, it was based upon and emphasised the dark colouring typically seen on the upper parts of animals' bodies to hide them from observation from above, and the light colouring on their underparts to hide them against the sky when viewed from below. For ships to emulate this effect, the advice was to paint all exposed, bright surfaces darker, while all under-surfaces, in shadow through lack of illumination, should be painted in a light tone. The result would be a flattening of the vessels' inherent structural pattern.

White was the colour recommended by Thayer for under-surface counter-shading purposes, but he went much further in regard to the application of this colour to ships. Following tests with cards in

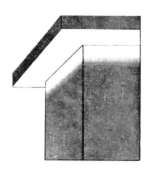

PAINTED ONE COLOR. This illustration shows the appearance of dark areas caused by overhanging deck when the entire section is painted one color. The underside is always in shade. Some shadow is cast on the bulkhead.

COUNTERSHADED. This illustration shows the same structure countershaded by the application of white paint in the dark areas. Undersides of all overhanging surfaces should be painted with white. The white should be brought down from the top on the bulkhead approximately one-third the amount of the overhang.

CAMOUFLAGE INSTRUCTIONS + Supplement to Ships – 2 + MARCH 1943   7

The Counter-Shading principle in an illustration from an official US Navy guide to ship camouflage. As early as 1896, Abbott Handerson Thayer had advocated the adoption of counter-shading as one of the best means of concealment for ships at sea. The principle involved deliberately painting bright, exposed surfaces dark, while areas in shadow, concealed by overhanging superstructure, should be painted white, the intention being to flatten the structural pattern. (US Navy Bu-C&R)

various grey tones, as well as black and white, suspended against a sky background, he suggested as part of his doctrine that white provided the most effective measure for concealment in all conditions against the horizon sky. This radical proposition was to induce a great deal of controversy, especially as empirical evidence clearly demonstrated otherwise. But that notwithstanding, Thayer's premise was not exactly for pure white. Rather, it had to contain a small measure of blue pigment. The resulting pale tint was called Thayer Blue, as adopted in the Second World War.

Whereas Thayer insisted that counter-shading was an all-embracing concealment mechanism employed by animals, there were clearly exceptions. Nonetheless, to deal with this aberration, while remaining steadfast to his belief that all animal colouring was intended for concealment (he chose to diminish its use for warning, or for attracting mates or pollinators), Thayer formulated a third concept which provided for when it was necessary for animals to be hidden against variegated backgrounds typical of the regions they inhabited. Typically, such animals were also marked in a patchwork of alternating dark and light areas in their coats in order to blend into their surroundings.

Identified as disruptive colouring or background-blending colouration, the essence of this alternative form of natural concealment was outlined as a universal 'Principle of Obliterative Coloration' in Chapters 3–9 of *Concealing Coloration in the Animal Kingdom*. The intention, it was explained, was that this, too, should govern the ways in which ships should be protected visually, a concept that was to be championed by another pioneer of maritime camouflage.

Thayer pursued the US Naval authorities with his manifesto, anxious for it to be taken advantage of, and later he made overtures to the British Admiralty, with an increasing intensity in August 1914, shortly after war broke out. In the UK, Thayer found he had a kindred spirit in Professor John Graham Kerr, a biologist and embryologist who was Regius Professor of Zoology at the University of Glasgow.

Though he became a devotee of Thayer's doctrines and was, more or less, his British counterpart, Kerr was certainly not Thayer's disciple, for he had in fact reached similar conclusions in parallel, based on his observations during field trips to Paraná on the River Plate, Argentina. As a result, he championed the same causes of counter-shading, which he called 'compensative shading', and disruptive colouration, which he named 'Parti-Colouring'. The latter was a critically important augmentation to Kerr's proposed means of diminishing visibility. The aim was, he declared, not only to conceal but to break up the recognisable lines of warships with randomly distributed, irregular bands or patches of white or light tone paint, in order to render fire control rangefinding more difficult. Parti-Colouring was broadly along the same lines as Thayer's thinking on obliterative colouration and Kerr also stressed the importance of implementing 'compensative shading', as he called it, along with the other components.

Like Thayer, Kerr approached the Admiralty to advocate the employment of his methods of camouflage by the Royal Navy. On 24 September 1914, not even two months after the UK had entered the war against Germany, he wrote to Winston Churchill, then the First Lord of the Admiralty, his covering memo supported by a dossier containing detailed descriptions of the camouflage practices he was proposing. Subsequent approaches, arising from frustration at the apparent unwillingness to fully adopt his scheme or the inability to completely grasp the nuances of its principles which, according to him, required precise adherence in the application process, ultimately amounted to something of a crusade which only elicited official irritation rather than a more favourable reaction.

The inference of this implied failure to act on his submissions as he desired was that little came of them when, in fact, quite the opposite was the case. Regrettably, Kerr's initial deposition was not fully and formally responded to by the Admiralty until July 1915, but in the meantime, in December 1914, he was unofficially notified that his ideas for protective ship painting had been communicated to the fleet for trial the previous 10 November, with the approval of Lord John Arbuthnot Fisher, the First Sea Lord. It had been disseminated in the form of a General Order under the title 'Visibility of Ships – Method of Diminishing'.

The first vessel to be so treated was the battlecruiser HMS *Indomitable* which, arising from its selection, was claimed (incorrectly) to have been the first naval vessel in the First World War to receive recognisable camouflage as distinct from an overall coat of plain grey paint. It is now difficult to identify all the Royal Navy vessels that were painted in accordance with Kerr's instructions, but it is believed, based on the study of contemporary photographs and by making reference to reports and research documents, that there were at least ten others, possibly more.

Revealed as being almost certainly the recipients of the Kerr scheme treatment were the battleships HMS *Irresistible*, *Canopus* and *Agamemnon*, the battlecruisers HMS *Inflexible* and *New Zealand*, as well as several cruisers and naval monitors. It would be reasonable to conclude that this amounted to a fair trial of significant magnitude, backed by generous Admiralty support. Certainly, most other camouflage schemes submitted for consideration at that time were not granted such a favourable response.

Just as he was to experience subsequently when he pursued the US Naval authorities with the same proposals, Abbott Handerson Thayer's overtures received short shrift by the Admiralty, even after he made a personal visit to London on 13 November 1915 to press his case. On the other hand, Kerr's concepts, though broadly identical to Thayer's, were given due regard and had the benefit of reasonably extensive evaluation, but, ultimately, they too were declined. The explanation given for the rejection, in an Admiralty communication of July 1915 from Captain Thomas E. Crease, Naval Assistant to the First Sea Lord, was plain enough. Drawing attention to the fact that constant changes to light level, the time of day, the position of the sun and the colour of the sea and sky all necessitated rapid adaptations to colouring schemes for them to be effective in a range of operational environments, the conclusion arrived at from the trials was '… the proposals are of academic interest but not of practical advantage'.

It should be borne in mind that this issue was not, by any means, the highest priority demanding Admiralty attention at that time. Embroiled in the ill-conceived Gallipoli Campaign, which to some extent gave rise to Churchill's replacement as First Lord of the Admiralty by Arthur Balfour, it is reasonable to conclude that the matter of this proposed ship camouflage method had been given adequate consideration in the circumstances. Nonetheless, the rejection was not well received by Professor Kerr, who contended that his painting instructions had not been correctly adhered to, a shortcoming which could have benefited from some practical advice and guidance. It was an oversight which, by his own admission, he heaped blame upon himself, '… it did not occur to me as necessary to insist that the application of these principles to be successful must be carried out under skilled supervision.'

Along with these instructional issues, there were also key logistical and material matters which were required to be addressed, just as much as the design details involved. This was especially important when the application of the scheme was intended across a vast fleet located in countless dockyards and anchorages at home and abroad. That aside, Kerr's concerns and disapproval of his official rebuttal were to escalate into a legal dispute that will be dealt with further on.

Although the camouflage concepts presented by Professor Graham Kerr were probably the first that were given thorough consideration under trial, they were not, as has been suggested, the first patterned schemes applied to Royal Naval vessels in the First World War. In the absence of official directives, numerous bespoke designs were adopted as an emergency expedient at the very outset of the conflict, each conceived and applied by individual ships' companies.

During a visit to Wester Ross, Ross and Cromarty, in September 1914, Winston Churchill had commented on the great variety of paint designs of 'queer mottled fashion' displayed on the ships in the anchorage. This was substantiated when, around that time, Commander Dudley Pound witnessed something along similar lines among the naval ships he observed anchored at Scapa Flow. These were all short-lived practices, though, and along with the rejection of Kerr's system (equally Thayer's concepts), the Admiralty's orders were for all ships to revert to standard Navy Grey. And, for British warships, that is how things remained, with very few exceptions, until certain naval units were painted in 'dazzle' colours from 1917.

However, for Graham Kerr, and indeed for Abbott Handerson Thayer, this was not regarded as the end of the matter. Both pursued the Admiralty relentlessly, seeking to get official policy changed, the tone of their communications becoming increasingly acerbic. At the same time, arising indirectly from another visit to the UK by Thayer in 1916, and prompted by Kerr, certain experiments into relative visibility

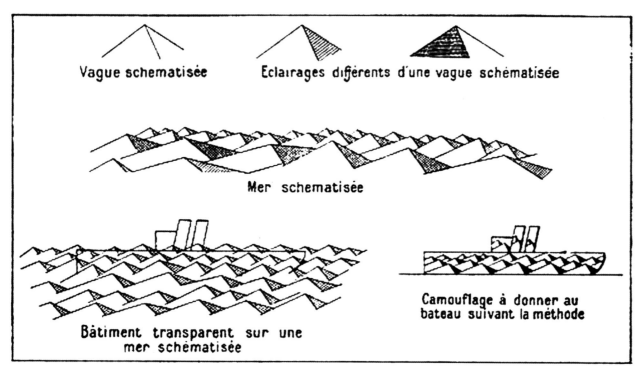

Vague schematisée   Eclairages différents d'une vague schématisée

Mer schematisée

Bâtiment transparent sur une mer schématisée

Camouflage à donner au bateau suivant la méthode

Typical of the early thinking in the development of ship camouflage was the concept depicted in this illustration, published in a book by Georges Clerc-Rampal in 1919, but dating from earlier in the First World War. It suggests a vessel would be less conspicuous if the appearance of the sea were painted all over it. These 'background-blending' ideas of simulated wave forms painted over ships' sides remained prominent well into 1917, advanced predominantly by well-intentioned amateurs. (Larousse)

This painting by C. McKnight Smith, dating from the First World War period, of a camouflaged ship with simulated sea decoration hints at the low-visibility scheme suggested by Zacarias Gomez, which also proposed the replication of a seascape on the side of a ship. (Roy Behrens)

Tested by the Eastman Kodak team under Loyd Ancile Jones, Gomez' scheme was not approved for use by ship owners. (Eastman Kodak)

were conducted between February and June 1916 by Sir Thomas Hudson Beare, Professor of Engineering at Heriot-Watt University, and the zoologist James Cossar Ewart, Regius Professor of Natural History at Edinburgh University, to determine how various tones of paint compared when viewed against the daytime sky. However, these experiments were not made in sea-going conditions, or even against a simulated sea backdrop in the laboratory, such that the principles they purported to illustrate were considered to be inconclusive and no further investigations were undertaken.

Abbott Henderson Thayer was a complex character, who tended to exhibit a rather pompous and dismissive manner in his dealings with official individuals and organisations. Inflexibly certain of the efficacy of his proposals, those examples that did not conveniently fit in with his thesis were somehow encompassed within its scope by convoluted reasoning and he exhibited an unwillingness to countenance any opposing view or consider any compromise or alternative. Attributing to himself a superiority in his understanding of camouflage matters, his comments were often intended to quell any form of counter-opinion. In *Concealing Coloration in the Animal Kingdom*, ostensibly written in 1909 by his son, he had assertively declared, 'Our book presents not theories, but revelations.'

Despite having identified important dimensions of natural camouflage where its application to ship camouflage was concerned, Thayer was inflexibly reluctant to consider the validity of cases when, for instance, the liberal use of white paint would be ineffective or counter-productive. The forceful and uncompromising promotion of his philosophies induced opposition, notably from the highest level, for it drew an equally vocal condemnation of his extreme claims from Theodore Roosevelt, the former president who, from 1913, was the US Assistant Secretary of the Navy.

Science advances on the basis of testing theories by gathering verifiable evidence in a controlled and unbiased way, avoiding foregone conclusions and having a willingness to adapt a concept as and when data suggests it is necessary. Thayer was not a scientist, but there is a question as to whether, apart from his artistic studies, he followed any form of rigorous data gathering to substantiate his theories. Nonetheless, despite its evident flaws, 'Thayer's Law of Countershading', as expressed within the 1909 publication, is widely regarded by biologists and remains accepted as being valid even today. To Thayer's credit, he is the only person involved with camouflage development to have had a natural law, a unique blue-white tint and a camouflage scheme named after him.

Though less adversarial as a protagonist, Professor Graham Kerr was also driven to get his message across. His personal papers in the Glasgow University Archives reveal that innumerable letters were written by him, pressing his case or eliciting support for his proposals. He generously volunteered his personal assistance to the Admiralty at his own expense and he also continued to pursue Winston Churchill, even though Churchill no longer held the position of First Lord of the Admiralty.

Like Thayer, Kerr had a tendency to convey his beliefs in a way intended to brook dissent or disagreement and, in so doing, he engendered opposition. Despite the fact that his proposals were based on his subjective deductions rather than the quantified results of controlled experimentation, in one of his letters he declared, 'It will be realised that what I have written above are mere statements of scientific fact.'

It has been suggested that the basic concept of Kerr's scheme has been misunderstood when, in reality, it is perfectly understandable. The premise appears to be that such a contention validates his scheme when it is contrasted with that of Norman Wilkinson (discussed in Chapter 3), against which it was to rival, whereas it should be judged on how effective it was in the achievement of its stated objectives. Kerr, himself, had used similar reasoning – that his scheme was not correctly comprehended – to thwart contradiction.

The salient point is that Kerr's proposals differed fundamentally from those of Wilkinson in that Kerr's system was designed exclusively for implementation on warships and particularly the big-gun units of the battle fleet. It was intended as a means of hampering enemy rangefinding through both reduced visibility and the breaking up of recognisable shapes and lines. Indeed, in May 1943, in the official Admiralty publication CB3098R, it is stated, 'The aim of these [Kerr's] measures was to reduce the visibility of ships at a distance and to hinder range-finding.'

There is no evidence to suggest that his proposals were ever intended, unlike dazzle painting, to interfere with judgement of course or speed, except as claimed after the event, in July 1919, in the 'Transactions of the North East Coast Institute of Engineers and Shipbuilders'. Equally, there is no suggestion that it was intended for application on troop transports or cargo ships as a means of countering submarine attack. No doubt, when senior naval personnel decided against Kerr's scheme, they based their conclusions on the feedback received from almost six months of practical trials. To state otherwise implies that Kerr knew better than experienced Royal Navy officers who had the benefit of years of sea time.

It should be kept in mind, as will be discussed later, that over the course of the First World War there were other attempts to develop means of hindering rangefinding against capital warships, none of which was successful. However, twenty or so years later, the background-blending and counter-shading elements of both Kerr's and Thayer's concepts were to enjoy a measure of revival and vindication, if not total acclaim, albeit in a questionable refutation of dazzle disruption, as the Admiralty attitude towards their ideas appeared to have turned full circle – or had it?

Meanwhile, in the USA, another maritime camouflage pioneer had emerged, William Andrew Mackay, a New York artist. He proved to be a leading light in the genesis of ship camouflage by approaching the matter of achieving low visibility from a completely different perspective, both in regard to his aims and in his application technique.

· CAMOVFLAGE · FOR · V.S.S. MOVNT · VERNON ·

Gerome Brush did not confine his camouflage efforts to counter-shading alone. This is his concept for the troop transport USS *Mount Vernon*, which exhibits pronounced disruptive elements comparable to the design features of Warner Dazzle and foreshortening to deceive range judgement. (US Naval History and Heritage Command, NH45766)

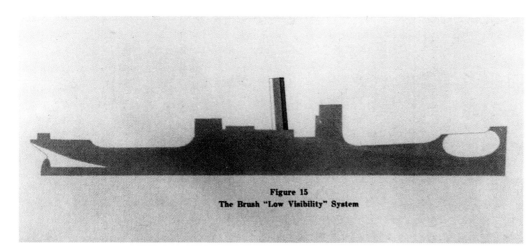

Figure 15
The Brush "Low Visibility" System

Opposite: Some Japanese warships from the First World War period were painted in a rudimentary type of camouflage. This is the battleship *Haruna* in c. 1916, bedecked in opposed chevrons on her funnels and stripes on the masts. It is not known how her hull was painted. (Shizuo Fukui)

Left: Side view of a ship painted in accordance with 'Thayer's Law of Counter-Shading' as submitted by Gerome Brush and tested by the Physics Department of Eastman Kodak at Rochester, New York. (Submarine Defense Association)

Before official camouflage measures were introduced, do-it-yourself schemes were the imperative early in the First World War. This is Monmouth-class cruiser HMS *Cornwall* in an early form of patterned camouflage. (Alamy Chronicle)

More Royal Navy ships in early unofficial camouflage schemes in late 1914/early 1915. Here the cruiser HMS *Dublin* is in a coat of white speckles over grey. (Imperial War Museum, Q73421)

HMS *Royalist* in a composite design, her hull painted in alternate areas of light and dark grey. The Royal Navy writer and engineer H.M. Le Fleming attested to the existence of these 'homegrown' schemes, noting that from 1914 to 1915 a number of ships were deceptively painted in mottled patterns of white and/or dark grey over a medium grey. (Imperial War Museum, SP2915)

HMS *Weymouth* seen in 1915, painted in the Kerr anti-rangefinding scheme. Characteristic of the scheme were irregular patches of white distributed randomly on the ship's hull and upper structures. (Alamy Historical Collection)

Among the small number of photographs that show naval ships painted in Professor John Graham Kerr's Parti-Colouring camouflage system is this view of the cruiser HMS *Argonaut*. Professor Kerr avoided prescribing background blending as part of his strategy, having qualified its effectiveness where large fleet units were concerned. However, he did imply benefits from the use of colours 'that were exact matches to the tint' of the background against which ships would be viewed. (Glasgow University Archives)

Painted in a similar fashion to Kerr's Parti-Colouring scheme is the French destroyer *Magon*, of the Bisson-class, early in the First World War. (Jean-Yves Brouard collection)

Oriflamme, a Branlebas-class ship, also in a similar scheme to Kerr's Parti-Colouring, although its alternating patches have hard edges. (Jean-Yves Brouard collection)

2

PAINTING WITH LIGHT – LOW VISIBILITY 2

As early as 1912, William Andrew Mackay had begun exploring mechanisms of optical deception arising from the chromatic interaction of different wavelengths of colour. This led him, with the aid of US Navy associates, to launch experiments that lasted from 1914 to 1917 in the concealment colouring on – of all things – submarines! Initially, the objective was to hide their extended periscopes but subsequently the trials progressed to concealment of hulls and conning towers.

Supported first by Commander Joseph O. Fisher, and later by Lieutenant Ronan C. Grady of the Office of the Chief of Naval Operations, combinations of green and purple stripes were tested on K-class submarines of the US Atlantic Fleet's 4th Submarine Division off Pensacola, Florida. Regular feedback was communicated to Josephus Daniels, the US Secretary of the Navy, but nothing came of the trials and the idea was eventually dropped.

Undaunted, Mackay, on his own initiative, adapted his New York studio at 345 East 33rd Street, Manhattan, into a camouflage training school – the first of its kind. Among his students were many of the artists who were to become prominent in the effort to develop and apply camouflage to US ships in the ensuing war and beyond.

The attention of Mackay and his team was primarily turned to the protective colouration of cargo vessels, converted troop transports and naval auxiliaries. He was mindful that, when, rather than if, the USA became a belligerent, these ship types would be in the vanguard, running the gauntlet of the waiting U-boats while conveying troops, armaments and supplies across the Atlantic to the battlefields in Europe.

William Andrew Mackay was pre-eminent among the embryonic camoufleurs in seeking to achieve an effect that would provide a measure of concealment in two situations – from both distant and closer observation. By exploiting the complex process of additive colour together with the limits of visual discrimination of detail beyond a blending distance, his aim was to convey chromatic differences in the apparent tone of a ship's colouring to assist reduced visibility in both circumstances.

Camoufleurs are interested by the way in which objects assume a colour (see image on page 11) and how the human eye perceives that colour. This process occurs in a myriad of wavelength combinations but, of importance, it also manifests itself in variations of grey tint when all three primary wavelengths of white light – blue, green and red – are present together to varying degrees. Moreover, as explained earlier, it can be seen that, as the constituent wavelengths of light energy emitted from the illuminating source change, so too will the perceived colour of the illuminated object.

It was these fundamental scientific principles that formed the basis of Mackay's camouflage endeavours. He sought to provide a painted design comprising elements of the primary colours that would simulate a grey, or grey-green tint equal to, but instead of, a single flat coat of grey paint of equivalent tonal value.

His method was to select shades of three primary colours or hues which he mixed together additively, using an adaptation of the principle of Isaac Newton's and James Clerk Maxwell's spinning discs, experimenting with variations in the amounts and blends of these colours until he found the combination that suited his purposes. Mackay's discs comprised a circular metal plate mounted on a spindle. The top surface of the plate was divided radially into sections upon which pigments of the primary colours – violet, green and red – were painted in the sequence of their changing wavelengths. If distributed equally, when the disc was rotated, the colours merged together and were perceived as being grey-white.

In Mackay's experiments, the surface area attributed to each of the primary colours of the spectrum as a percentage of the whole disc was repeatedly adjusted in order to create a composite bias towards a particular grey tone. The ratios of the three colours which finally combined to match the desired tone of his choosing became the surface area proportions of the paint pigments to feature in his camouflage patterns, one such mixture having been noted as violet 44 per cent, green 20 per cent and red 36 per cent. At a distance, the colours would merge into the single tinted-grey tone Mackay sought, whereas close up, the discernible pattern could cause a measure of deception. Also, as daylight changed during the course of the day from a red cast through blue and back to red, along with the influence of the Purkinje Shift on the perceived spectral frequency of the green paint present, the apparent colour tone of his camouflage scheme could change to appear lighter or darker.

Using a technique similar to that of the pointillistic style of French post-impressionist painter Georges Seurat, application of the selected colour pigments was in small, speckled patches of two of the colours juxtaposed over a third base colour. Typically, they were painted as red and green speckles over a violet purple base. When asked why not just mix the green, red and purple pigments together in a bucket, Mackay replied, 'Because I am not merely dealing with paint. It is light I am mixing.'

The first ship to be painted in Mackay's Low-Visibility camouflage scheme was the auxiliary USS *Olean*, but it was not long before it was applied to numerous other ships, among them converted troop transports, escort destroyers and other naval auxiliaries, being among the first of the US ship camouflage systems to be officially approved. Later, Mackay patented his scheme under the title 'Process of Rendering Objects Less Visible Against Backgrounds' (Patent No. 1,305,296, 3 June 1919).

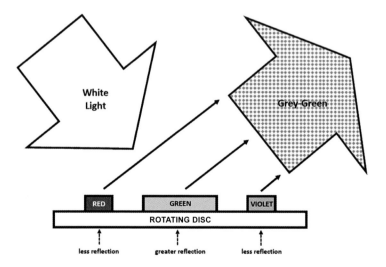

A simplified diagram illustrating Mackay's colour-mixing principle. Grey tones are seen when all wavelengths of white light are present but reflected incompletely in varying percentages: a higher percentage for light grey, a lower percentage for dark grey. A bias to a particular wavelength – in this example, green – would tint or shade the grey tone. In practice, in Mackay's system, the percentages were arrived at by varying the areas or the concentration of speckles of each of the primary colours. Thus, when the eye was visually unable to resolve the detail of the speckled pattern, it would blend, in this example, as grey-green. (Author)

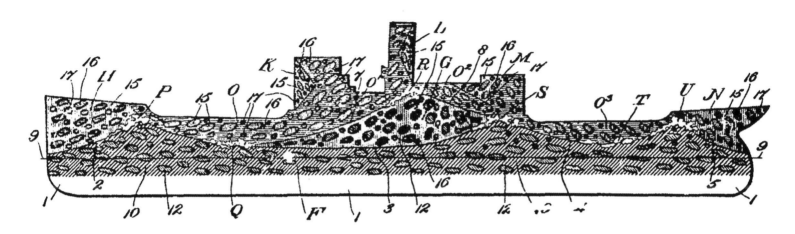

A drawing of Mackay's patented Low-Visibility system of ship camouflage, showing in detail its constituent elements. As applied in practice, the design consisted of two distinct areas, each comprising a mixture of primary colours of different combined tonal value, although three such areas are indicated in this patent drawing. The lighter tonal combination was painted uppermost while the darker combination extended downwards beyond the waterline. The areas of pattern were separated by an undulating dividing line along the side of the hull intended to simulate the peaks and troughs of a wave formation such as might be seen in an active sea state, its purpose being to assist background concealment. (US Patent Office)

William Andrew Mackay seen viewing a model ship through a periscope. Ironically, the model is not painted in one of Mackay's schemes but has a dazzle pattern. (Roy Behrens)

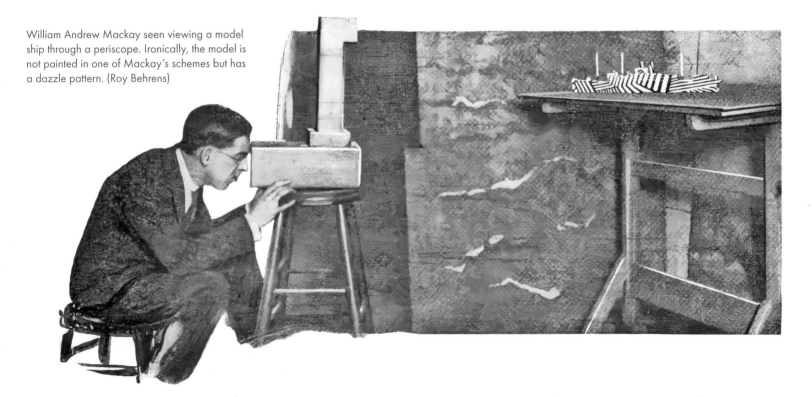

As anticipated, the US entered the war against Germany on 6 April 1917, the continuing provocation of the U-boat offensive against US and other neutral shipping having acted in part as the catalyst. Within days of the USA declaring war, the Naval Consulting Board ramped up its involvement in the pursuit of camouflage protection for US ships. The Naval Consulting Board, a quasi-governmental institution briefed with pursuing all technological initiatives to enhance the performance of the US Navy, had been created on 13 September 1915 by Secretary of the Navy Josephus Daniels, on the suggestion of the US inventor Thomas Alva Edison, who became the board's first president.

Naturally, the Naval Consulting Board had close connections with the Navy Department and also, after it was established on 7 September 1916, with the US Shipping Board, the body responsible for merchant shipping. The following year, on 16 April 1917, the Shipping Board set up the Emergency Fleet Corporation under General Manager Major George W. Goethals, which, in conjunction with the Naval Consulting Board, then instituted the Ship Protection Committee, headed by Rear Admiral Harry H. Rousseau.

These were key developments for US camouflage initiatives. With both the Naval Consulting Board and the Emergency Fleet Corporation represented on the committee, it brought the interests of naval and merchant shipping protection under a common umbrella.

Arising from these steps and at the behest of the Ship Protection Committee, on 27 June 1917 the Navy Department authorised trials of different camouflage schemes applied to nine ships. Two were to be painted in standard Navy Grey, two in Mackay's Low-Visibility scheme and two each in the Herzog and Toch Low-Visibility Dazzle schemes (described in Chapter 5), the latter selected from a number of privately developed camouflage concepts submitted for consideration from artists, industrial chemists and paint manufacturers.

The ninth ship was painted in the Brush Counter-Shading scheme. Submitted by Gerome Brush, this was ostensibly the concept originally promulgated by Abbott Handerson Thayer, Brush having been the co-registrant with Thayer of the 1909 patent.

Subsequently, on 1 October 1917, on the recommendation of the Ship Protection Committee, the US Treasury Department, through its Bureau of War Risk Insurance, notified ship owners that they were required from that date to paint their vessels in one of the approved protective measures or, if they failed to comply, face a penalty of an increase in war insurance premiums of 143 per cent. The regulation stated:

Each vessel shall be painted in accordance with one of the systems that are recommended by the Chairman of the Naval Consulting Board and the Ship Protection Committee of the United States Shipping Board. The following systems have been approved to date: William A. Mackay, Everett L. Warner, Maximilian Toch, Jerome Brush and Louis Herzog.

It is to be understood that ship owners are free to select any one of the approved methods for their own use. Should a ship owner desire to follow his own method, it must first be submitted to and receive the approval of the Chairman of the Naval Consulting Board. Upon completion, the ship owner must furnish the United States Shipping Board inspector, or his representative at the loading ports with a certificate …

The addition of the Warner scheme (discussed in Chapter 4) to the approved list is interesting insofar as it was the only approved scheme that had not featured in the Navy Department trials.

While adoption of these measures was encouraged, no evidence has been found of guidance as to the practicalities of accomplishment – how the necessary supplies were to be sourced and prepared, how dockyards and ship maintenance bases were to be engaged to carry out the work and, the thorniest question, who was to pay for the costs involved. Undoubtedly, something of the kind must have been promulgated – now lost in the dungeons of official archives – for ships so painted were soon evident. It was reported that, by December 1917, there were numerous examples of ships painted according to the Brush, Mackay and Toch schemes, with somewhat fewer in the Herzog camouflage design.

More or less concurrent with the actions of the various US governmental bodies, a private initiative was launched by a group representing almost 100 leading merchant shipping lines. The Submarine Defense Association was created in May 1917, with offices at 141 Broadway, New York, with Lucius H. Beers, attorney for the Cunard Line, as its chairman. Other executive officers included John A.H. Hopkins as vice chairman and Lindell T. Bates as secretary treasurer.

The son of the latter, Lindon W. Bates, was engaged to chair the association's Engineering Committee. Various specialist advisers were recruited for their expert knowledge, principally of visibility and detection, among them one Harold Van Buskirk. Besides the Cunard Line, other British shipping companies that joined the association were Anchor Line, Furness, Withy & Co., Lamport & Holt and Bowring & Co.

After its establishment, the Submarine Defense Association wasted no time in embarking upon a number of important initiatives. Foremost, by August 1917, it had made an arrangement with the Eastman Kodak Company at Rochester, New York, whereby the Physics Department in its research laboratories, under the leadership of Loyd Ancile Jones, would conduct thorough evaluations of the effectiveness of the various camouflage schemes then under consideration by the Ship Protection Committee for approved official adoption.

Loyd Jones was a leading physicist of the day, known and respected internationally for his work on photometry, sensitometry and colorimetry. In the field of photographic film speed measurement,

his research would ultimately lead to the ASA (American Standards Association) system of film rating, now referred to as ISO.

In customary fashion, Jones set about rigorously formulating means of reliably measuring visibility, designing unique tools for that purpose, as well as apparatus for viewing models of ships painted according to the already prescribed camouflage measures. His declared objective was to seek the means by which 'the claims of the various [proposed] systems might be correctly evaluated by a method not involving personal judgement'. The most extraordinary feature of the complex indoor observation theatre he had specially commissioned for the task was its sheer scale. Comprising a shallow water tank, 14ft in diameter, under a large diffusing dome, the means of observation of models was by a periscope optical system mounted inside a motorised, wheeled enclosure described as a 'truck'. This truck travelled on a rail track, 130ft in length, inscribed with scales in yards and knots.

The extent of Jones' scientific developments in the course of these studies of camouflage systems cannot be overstated. He designed a visibility meter and a colorimeter, and he devised numerous

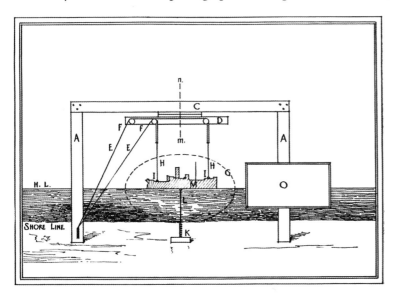

Besides an in-house model-testing theatre, Jones also set up a viewing rig on the shores of Lake Ontario, upon which cut-out shapes of ships could be viewed against a natural sea and sky background, as shown in this drawing. (Submarine Defense Association)

A view of the elaborate model-viewing facility at Rochester. Here, a model is seen through the aperture which was designed to simulate the restricted field of vision experienced when observing through a submarine periscope. Note the rails in the foreground for the viewing apparatus to travel along. (Submarine Defense Association)

Looking beyond the aperture of the model-viewing facility constructed by the Loyd Jones team, this view shows the diffusing dome structure above the water tank, by which illumination of models could be adjusted and controlled. (Submarine Defense Association)

mathematical formulae and relevant optical terms to permit accurate measurement of such things as the functional relationships between visibility and the weather coefficient, visibility and the reflection factor, the sensibility of the eye to brightness contrast, etc. Ultimately, his team assisted in the conception of two camouflage practices for the Submarine Defense Association, based upon the application of specially formulated paint pigments, each of which was identified by a Greek alphabetical character.

A strict adherent of the low-visibility school of marine camouflage thinking, Loyd Jones applied that criteria in the visibility evaluations he conducted on the camouflage schemes so far tendered for deliberation, disregarding any supplementary or alternative perceptual interference quality they may have exhibited which could assist in the protection of ships from attack, even though such characteristics could increase their visual prominence. This bias was evident in his published table of the comparative degrees of visibility.

The schemes that were assessed consisted of the four then undergoing Navy Department trials plus other proposals submitted by the artists Walter Karl Pleuthner and Zacarias Gomez and the paint-manufacturing firms Sherwin-Williams and Patterson-Sargent. Also evaluated were British Dazzle (see Chapter 3), Warner Dazzle, Mackay Low-Visibility Dazzle and two light grey tones devised by the Jones team from which specific Psi and Omega greys were subsequently contrived. The tests were based on uniform simulations of a 6,000-yard viewing distance in atmospheric conditions typically encountered in the North Atlantic submarine danger zone. A rating of 0.0 would have indicated total invisibility. The results are shown in the table on page 35. It is interesting to note that, although the two light grey tones registered the lowest visibility ratings, five schemes of the dazzle or low-visibility dazzle type performed better than those intended to achieve low visibility alone.

Due recognition for the first awareness, in the context of marine camouflage, of the implications of differing atmospheric conditions across oceanic regions should rightly be attributed to Abbott Handerson Thayer, for he had outlined the effect of these differences in his proposals. But the Jones team at Eastman Kodak went further. Besides deducing the general meteorological conditions that prevailed

A third view of the model-viewing facility at Rochester. The viewing apparatus, mounted inside the light-tight cabin or 'truck' shown here, travelled back and forth along the rails. Jones' scientific approach to the issues associated with ship camouflage development resulted in numerous novel algorithms for calculating various interrelational aspects of light, colour and visibility. (Submarine Defense Association)

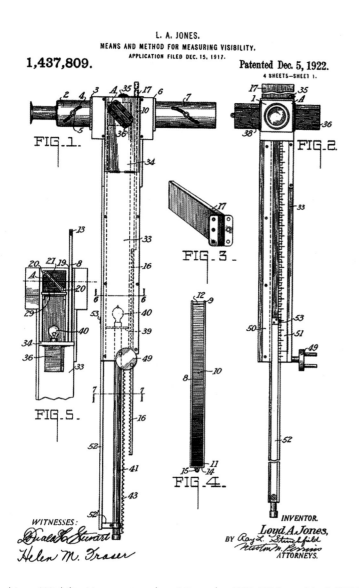

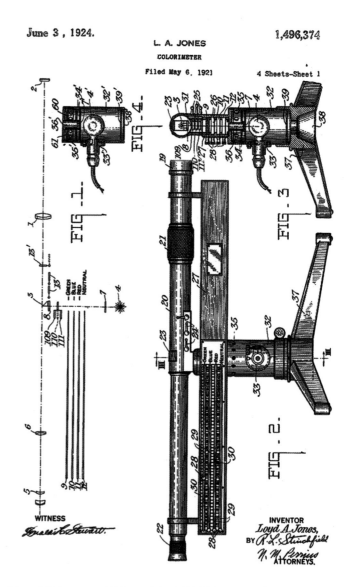

Loyd Jones' Visibility Meter, patented on 5 December 1922, US Patent No. 1,437,809. (US Patent Office)

Loyd Jones' Colorimeter, US Patent No. 1,496,374, registered on 3 June 1924. (US Patent Office)

Model	System	Weather Coefficient	Visibility
F 20	Light Grey	.43	0.3
F 9	" "	.42	0.6
Herzog	Low Visibility-Dazzle	.43	4.7
Mackay	" " "	.42	4.4
Toch	" " "	.42	6.0
Pleuthner	Dazzle	.43	7.2
Warner	"	.42	7.2
Gomez	Low Visibility	.43	7.5
Brush	" "	.43	7.5
Mackay	" "	.42	8.2
Patterson-Sargent Company	Dazzle	.43	9.5
Sherwin-Williams Company	Low Visibility	.43	10.0
British	Dazzle	.42	14.0
F 6	Black	.42	24.0

Table of comparative visibility ratings from the observation trials conducted by the Eastman Kodak Physics Department. (Submarine Defense Association)

Another low-visibility scheme assessed by the Eastman Kodak Physics Department which failed to gain approval for its use was this concept by Walter Karl Pleuthner. In its granular composition, it followed the logic of Mackay in consisting of an equally fragmented coat of multiple-coloured elements. (Eastman Kodak)

throughout the year for those sea areas in which the majority of ships were being sunk by U-boats – the eastern North Atlantic and sea approaches to the UK – they also meticulously analysed weather reports and mapped the results.

Arising from this, the first camouflage schemes emerged, categorically assigned for use and recommended for implementation in distinct climatic zones. In Loyd Jones' words, 'These designs take into account the average weather conditions in the various sections of the danger zone.' They were:

- The Low-Visibility Deception System for Northern Latitudes – where for 70 per cent of the time the weather was cloudy. The colour to be used was designated as Omega Grey, but alternatively the colours Alpha Blue and Beta White could be applied in equal proportions, such that by the correct juxtaposition they would blend into Omega Grey at the requisite distance.
- The Low-Visibility Deception System for Southern Latitudes – sea areas where far less cloud cover was expected. For this variant, the designated overall colour was Psi Grey or the alternative combination, again applied in equal proportions, of Gamma Blue and Delta White.

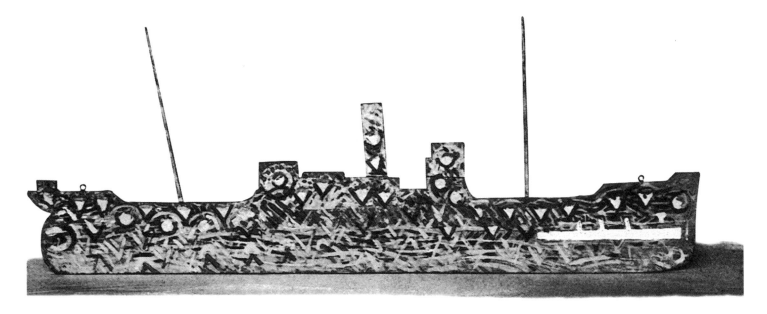

Both systems were characterised by muted disruptive patterns. The painting of false bow waves or a false bow was also advocated. In both cases, they were designed to cause course, speed and distance deception when viewed at ranges within 1,000 yards, while at a distance of approximately 5,000 yards and beyond the effect would be low visibility.

In an oblique concession to Abbott Handerson Thayer, the Submarine Defense Association schemes proposed combinations of whites and blue-greys, colours that were broadly similar to those that Thayer had insisted were the most effective for inducing low-visibility effects.

Armed with these formulations, a deputation from the Submarine Defense Association, led by Lindon W. Bates, met with Admiral William S. Benson, Chief of US Naval Operations, in September 1917 to present their findings and elicit his support to enable full-scale sea trials to confirm the efficacy of the two systems. The requisitioned steam yacht USS *Gem* was duly allocated and trials commenced on 12 December 1917, continuing through to 10 March 1918.

On her port side, USS *Gem*'s hull was painted overall in Omega Grey. On the starboard side, she had alternate horizontal bands of Alpha Blue and Beta White with a diamond-shaped pattern at the bow in the same colours. Some observations were made from the deck of a submarine, others from points ashore, the atmospheric conditions being mostly hazy and foggy. It was recorded that, with normal, unaided vision the chequered patterning blended at 1,100 yards and the coloured bands merged together at 2,500 yards.

These efforts culminated in the generation of a report by the Submarine Defense Association, written in two parts, the first by Lindell T. Bates, 'The Science of Low Visibility and Deception as an Aid to the Defense of Vessels Against Attack by Submarines', and the second by Loyd Ancile Jones, 'Protective Coloration as a Means of Defense Against Submarines', published on 8 March 1918. In its preamble, it states, 'That the staff of the Association be authorized and directed to deliver to the new Camouflage Section of the US Navy Department, as requested, all the Association's data concerning camouflage.'

This signified the termination of the activities of the Submarine Defense Association, which was wound down from that date, the baton having been passed to a new Naval Camouflage Section that was already in the process of being set up by Admiral Benson before even he had read the Bates–Jones report. That Camouflage Section came into being on 15 March 1918, organised into two subsections, one at the Eastman Kodak laboratories, Rochester, the other in the offices of the navy, at Pennsylvania Avenue, Washington DC.

Across the Atlantic in London, the Admiralty had by then already set up its own Dazzle Section, housed in Burlington House, Piccadilly, the home of the Royal Academy. Whereas the motivation in the USA had been, to a significant extent, to enhance understanding of colour and visibility behaviour as a means of scientifically advancing ship camouflage practices, the motivation in the UK was the urgent and critical need to counteract the worsening shipping losses following Germany's renewed policy of unrestricted submarine warfare. Launched on 1 February 1917, it threatened dire consequences for the nation.

Coordinated under a single authority, the progress of US camouflage development gathered momentum with its common focus. It was to remain in the vanguard for the foreseeable future, through to the end of the Second World War. Although it functioned at a reduced level in the interim, it was sustained within the framework of the US Navy as an important dimension of naval tactical research. In contrast, treated more as a short-term expedient, the British effort was dissolved once victory had been secured. Not even dormant, it had to be launched anew twenty years later when national survival was once more threatened.

USS *Olean*, completed in 1917, was the first ship to be painted in accordance with the Mackay Low-Visibility scheme. (US Naval History and Heritage Command, NH65090)

The transport USS *Minnesota*, formerly the Atlantic Transport Line ship of the same name, painted in Mackay Low-Visibility. (US National Archives, 19-N-7569)

Red Star liner *Finland*. The weathered paint on her hull could have undermined the effectiveness of the Mackay Low-Visibility scheme. (US National Archives, 165-WW-70G-004)

Also in the Mackay Low-Visibility scheme is the Caldwell-class destroyer USS *Conner*. (US Naval History and Heritage Command, NH70857)

Clockwise from top left: The wooden-hulled converted fishing vessel USS *Winfield S. Cahill* painted in the Mackay Low-Visibility scheme, photographed at the Norfolk Navy Yard, Virginia, on 18 August 1917. (US Naval History and Heritage Command, NH75515) The requisitioned steam yacht USS *Gem*, seen with elements of the Submarine Defense Association's Northern Latitude Design painted on her starboard side. Trials with the vessel took place in Long Island Sound, south of New London, Connecticut, and in the waters between Fisher's Island and Montauk Point, continuing through to 10 March 1918. (Submarine Defense Association); The converted passenger liner USS *Knoxville*, previously *St Paul*, another example of the Mackay Low-Visibility scheme. (Roy Behrens)

The destroyer USS *Caldwell* off Mare Island Navy Yard, California, on 14 December 1917. (US Naval History and Heritage Command, NH70850)

3

SEEING IS BELIEVING – BRITISH AND FRENCH DAZZLE PRACTICES

The diametrically opposed school of thought to low visibility is disruptive, confusion or deception painting. In his section of the Submarine Defense Association's report of 8 March 1918, Loyd Ancile Jones succinctly described this conflict between the two: 'The subject of protective coloration [for ships] quite naturally divides itself into two main divisions: "low visibility" coloration and "deceptive" coloration. These in general are antagonistic in principle.' Like the low-visibility measures, though, disruptive painting took more than one form, quite apart from the fact that it also involved countless unique designs and patterns.

As a concept, disruptive or 'obliterative' colouration had been introduced by Abbott Handerson Thayer and John Graham Kerr, as outlined earlier, but where their objectives had been to break up the structural lines of warships to hinder rangefinding, the aim of the disruptive painting that emerged in the spring of 1917 was quite different. Its purpose was instead to make it difficult for submarines to launch successful attacks against merchant ships by the use of optical illusions to interfere with judgement of target distance, speed and, most importantly, bearing.

If the definition of disruption in the context of camouflage is elucidated, these differences are manifested more clearly in the word's alternative meanings:

a) Visual disintegration or division into smaller cognitive or visual units, each of which matches a part of the background, leading to confusion as to an object's presence or shape.
b) Visual disturbance or interference which impedes an activity or process, tending to cause confusion as to an object's distance, size and, if moving, speed of movement and whether approaching or receding.

Definition (a) refers to the aims of Thayer and Kerr, whereas definition (b) refers to the principal objectives of the disruptive camouflage exponents of the later years of the First World War.

The best-known manifestation of disruptive painting of the latter type was 'dazzle', an extreme variant of deceptive colouration. Disruptive painting has been loosely categorised as being obtrusive (pronounced and stark) or unobtrusive (muted or less strident).

Typical of obtrusive disruption are elaborate visual effects manifested in a combination of dissonant, contrasting colours and bold, geometric patterns. But there was more to it than just randomly painting vivid, psychedelic colours and patterns on ships' sides. The colours and patterns selected had to achieve particular effects, such that, in their development, there was a requirement for evaluation of the factors involved and the careful testing of conceptual designs under simulated conditions.

Disruptive painting was controversial because, to achieve its objectives, not only did it fail to contribute to a reduction of visibility but it actually did the opposite – it tended to make ships significantly more visible. Its exponents argued that low visibility was, in practice, unachievable because, however a ship was painted, the background against which it was viewed was forever changing and required the colouring to be constantly adapted to suit – an impossibility. Moreover, in clear conditions, a ship would be plainly visible, its presence revealed at a considerable distance, even beyond the horizon, by the smoke from its funnels. The increased use of hydrophone equipment by U-boats also facilitated the detection of target vessels when they were still a good way off. For the disruption school, the real issue was not how to hide a ship but how to deal with the threat once the target was close enough to attack. As Norman Wilkinson said, 'Since it is impossible to hide a vessel, it does not matter how visible she is providing her course remains a matter of question to the attacker.'

Regardless of its evident flaw, disruptive painting, it was argued, offered a more functional solution because potentially it remained an effective deterrent at all times, since there would always be some part of its colouring that would be lost against the background, irrespective of the time of day, the atmospheric conditions or sea state. If the factors of primary concern for low-visibility objectives had been the nature of visibility and the quality of light in the battlezone, for disruptive painting they were an appreciation of the mechanisms of visual perception and an understanding of not only how U-boats launched their attacks but also the preparations they underwent preceding the launching of their torpedoes.

Although his remarks were expressed in the context of ship visibility, Lindell T. Bates appositely reiterated this vital importance in the Submarine Defense Association Report of 8 March 1918: 'A practical study of the scientific aspects of the visibility of ships requires not only a technical knowledge of the physics of light, color, and optics in general, but also an intimate familiarity with all the elements of submarine warfare. This familiarity is imperative.'

Cognisance of the capabilities of U-boats and the attack tactics developed for Germany's undersea fleet were equally fundamental to the strategy of disruptive painting, in order to identify any weaknesses that could be exploited. Knowing U-boat limitations, their preferred attack distance and the speed and range of torpedoes, if ascertainable, would all prove helpful to the development of camouflage countermeasures.

The U-boat, hidden beneath the sea with its lethal weapons, undoubtedly constituted a grave menace but, on the outbreak of the First World War, submarines were still fairly rudimentary compared to their later counterparts. They were slow (though faster on the surface than most merchant cargo vessels of the day) and generally small in size, having cramped interiors and a restricted weapon payload. The amount of time for which they could remain submerged was limited and while on the surface they were vulnerable to attack from destroyers.

Similarly, torpedoes of the type carried during the First World War also had their limitations. Due to the confinements of space, the U-boats were obliged to carry a type that were smaller and older, having a slower speed and a shorter range. These torpedoes were of the straight-running type, fitted with a gyroscopically controlled steering system which required course settings to be calculated and pre-set using a triangulation method. This overcame the need for the submarine itself to be steered onto the bearing on which the weapons were to be released, but the calculation of the gyro angle, dependent on a number of variables, all determined by visual judgement – a process typically repeated several times in the lead-up to the attack – was time-consuming and prone to error. In the absence of automated calculation devices, the computation of the angle of torpedo attack required to obtain a successful intercept with the target was a manual process.

The fire control crews at that time had only basic torpedo direction calculators in the form of rotating angle solvers and slide rules to assist them in their mathematical reckonings.

A crucial dimension of the pre-attack sequence which had a significant bearing on the accuracy of determination of the target parameters was the U-boat's means of observation on which the estimation of all three target variables – distance, speed and bearing – totally depended. First World War U-boats were equipped with stereoscopic rangefinding periscopes in which split images of the target were aligned, allowing the distance to be read off a micrometer scale to determine target range. This was not easy to do when the U-boat and target were in constant motion in a seaway, with the viewing system also subject to visual interference from waves and spray at sea level.

A further complication was that periscope optics, like all complex optical systems, suffered light loss by absorption and internal reflection. Typically, this would be compensated for by pupil dilation, but its consequence could be operator fatigue. The difficulties in rangefinding from a submerged submarine were explicitly spelt out by a contemporary periscope expert, Dr F. Weidert: 'Now it is already well enough known that correct range estimation with the naked eye without some means of assistance is extremely difficult and is for many people actually impossible. With … vision through an optical instrument this is even much more the case.'

The Submarine Defense Association Report devoted a lot of attention to this matter, placing great emphasis on the need to have a clear understanding of submarine instrumentation, how target characteristics were assimilated and the overall attack methodology followed. As revealed at the time, the odds were not entirely with the U-boat, 'Because the [target] data for the most part can only be calculated with difficulty, or not at all, it has to be estimated'. For the camoufleurs who favoured disruptive painting practices, such a revelation reinforced their case that it was possible to exploit the difficulties and hamper accurate target estimation.

The determination of target speed and course was dependent on the initial estimation of range which, as stated above, was known to be prone to inaccuracy. Thus, where speed and course estimation were concerned there was further potential for deception using camouflage.

An approximation of the target ship's bearing was deduced, relative to the submarine's course, by making comparisons each time the vessel was observed. Hardly scientific in its technique, it was complicated further if the target vessel was on a zig-zag or 'S' course, constantly changing its bearing, deviations that were not easily discernible from periscope level. While alignment of masts, funnels and other recognisable features could assist the submarine commander in making his assessment, his judgement could be undermined by the perspective distortion of disruptive camouflage.

Likewise, judgement of the target's speed was also conditional on the accuracy of the range calculation as well as an approximation of target size if sense was to be made of either the time it took for the vessel to pass between cross-hairs in the periscope optics or the unfolding measurements of the angles subtended to the target from the submarine. If it was assumed that a zig-zag course was being followed, some retardation allowance had to be made for the duration of the ship's passage between two points. In order to assist these estimations, U-boat commanders were known, on occasion, to take bow-wave indications into consideration.

The procedure typically followed in the lead-up to an attack, as recommended to officers by the fleet authorities, involved making initial observations on the surface at 15,000–20,000m, continuing to do this until there was a danger that the U-boat could itself be sighted from the target. Thereafter, the submarine submerged and the remainder of the attack process took place below the surface. At intervals, further observations were made, each lasting from five to thirty seconds in duration, each time adjusting the gyro setting. Meanwhile, the U-boat manoeuvred into the firing position, ideally abreast or marginally ahead of the target, running on a parallel course at a range of around 1,000–2,000m.

Early in the First World War, U-boats had generally attacked on the surface, often as close as 300–600m, using their deck guns in order to preserve their precious few torpedoes. In the face of improving countermeasures, attacks were increasingly executed while submerged and, to keep safe from naval escorts, the distance at which attacks were launched tended to increase up to and beyond 1,000m. This was well within the range capabilities of the torpedoes then carried but there was

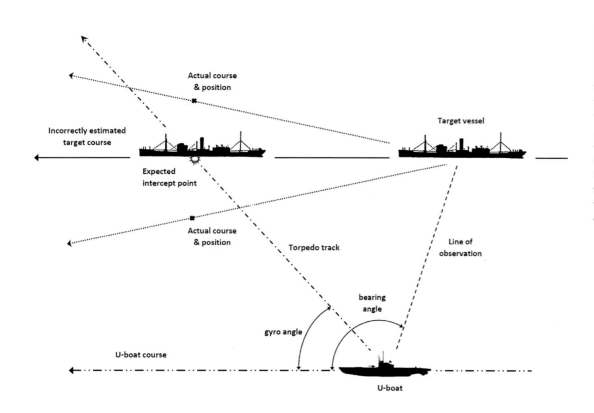

Diagram illustrating the basic principle of Dazzle as a means of course confusion. It shows the camouflaged target as seen and its projected strike position, the U-boat position, the target bearing, the gyro angle and the torpedo track. The angled lines subtended from the target's seen position show its potential course tracks on either side of the anticipated track, should the target bearing be misjudged through the confusion caused by the disruptive design. In these cases, the torpedo would either pass ahead of the target vessel or behind it, in either case missing it altogether. (Author)

a direct correlation between the range at which the attack was carried out and the likelihood or otherwise of a successful outcome. The greater the range, the greater the chance of a torpedo miss through even a minor discrepancy in the calculation of the gyro angle.

All in all, the U-boats confronted numerous impediments and the undersea war was not as one-sided as might be supposed but, despite all these obstacles, the U-boat commanders honed their attack skills and their proficiency was already at a high level before the decision was made, on 1 February 1917, to revert to unrestricted submarine warfare, essentially attack without warning. The impact of this on shipping losses was immediate and alarming. A projection of the sinkings by quarter, from the outset of the war up to the end of 1917, starkly reveals the grave situation that was materialising for the Allies, and the UK in particular. A greater number of ships and tonnage was lost – sunk or critically damaged – in the first six months of 1917 (2,326) than had been lost in all the previous twenty-eight months since the war had started (2,290).

	First Qtr	Second Qtr	Third Qtr	Fourth Qtr	Total
1914	–	–	–	15	15
1915	61	229	299	167	756
1916	174	252	451	642	1,519
1917	963	1,363	798	600	3,724

Such losses were unsustainable and spelt disaster, prompting a commensurate response in the form of a series of countermeasures, in the case of the UK encompassed within amendments to the Defence of the Realm Act of 1914. First and foremost, the convoy system was revived, an old practice that had been abandoned in the Victorian era, whereby a number of ships sailed in a group together, protected by an ocean-going naval escort. Anti-submarine patrols commenced and efforts were made to increase naval aviation capabilities to provide air cover, but realistically this was not available to any great extent, even by the war's end.

Significantly, another stratagem adopted was a radically different form of ship camouflage – disruption. Unlike in the USA, pursuit of low-visibility measures had not progressed in the UK beyond the proposals of John Graham Kerr, which had been rejected, and there was no evidence to support dependence on plain Navy Grey. Indeed, the latter was described by one commentator as 'disastrously inadequate'. Step forward a naval lieutenant commander with his thoroughly counter-intuitive solution, at odds with the received wisdom of the day and in conflict with all camouflage philosophy prior to that time.

He was Lieutenant Commander Norman Wilkinson RNVR, master of a naval minesweeper, an experienced yachtsman and an established marine artist who had amassed a wealth of understanding of the behaviour of light, colour and weather at sea from the studies for his canvases. Wilkinson was convinced that to protect ships by trying to hide them in a sea and sky background was an impossibility and, after having personally observed conspicuous, black-hulled transports at Gallipoli, he saw there was a compelling requirement for something more radical if torpedo attacks were to be hindered.

Drawing on his artistic background and mindful that obtaining critically important distance, speed and bearing parameters was fraught with difficulties, he believed such judgement could be undermined further by using geometric patterns and designs painted in contrasting colours to distort appearance and interfere with visual perception. Arcs and curves would break up structure; lines, angles and other geometric shapes and texture patterns could falsify perspective; and the apparent positions of bows and other features could be altered using contrasting paint colours, in combination creating confusion in calculating essential gyroscope settings.

Naming the concept 'Dazzle', Wilkinson arranged a visit with Captain Charles Thorpe, Flag Captain at Devonport Dockyard, on 27 April 1917, to present the essence of his proposals and request approval for them to be put to the test on a full-scale ship. Whether or not it was fortuitous timing, the U-boat threat then being at its height, his submission received a positive response, in marked contrast to the reception that Kerr's persistence had seemingly elicited. Initially,

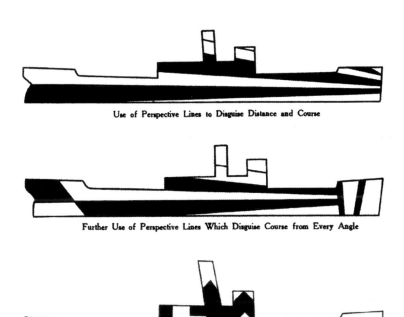

Use of Perspective Lines to Disguise Distance and Course

Further Use of Perspective Lines Which Disguise Course from Every Angle

Ambiguous Dazzle Amidships Conceals Direction of Course

Mechanisms of perspective distortion in dazzle painting. The top figure shows the use of perspective lines to interfere with estimation of course and distance. The middle figure illustrates a design intended to distort course estimation from every angle. The bottom figure shows the application of ambiguous dazzle patterning amidships to further confuse course direction. (Alon Bement)

Wilkinson sought the use of a single ship to permit trial observations of his concept, coordinated with naval units stationed ashore.

On 23 May 1917, the Admiralty gave Wilkinson's scheme its blessing and approved his application for a trial vessel. The Royal Fleet Auxiliary storeship *Industry* was duly allocated and, after it had been swiftly repainted, it embarked upon a coastwise voyage. The feedback was extraordinarily positive, and within days the painting of fifty more ships was authorised, each with a different dazzle pattern.

Work of such magnitude, with a necessity to avoid repetition, required adequate research and design facilities and sufficient staff to run them. Despite the endorsement of the Admiralty, suitable premises for the operation were not immediately forthcoming, so Wilkinson took it upon himself to approach the Royal Academy in London, where his

OBTRUSIVE COURSE DISTORTION DESIGNS. CRUISER. BROADSIDE VIEWS.

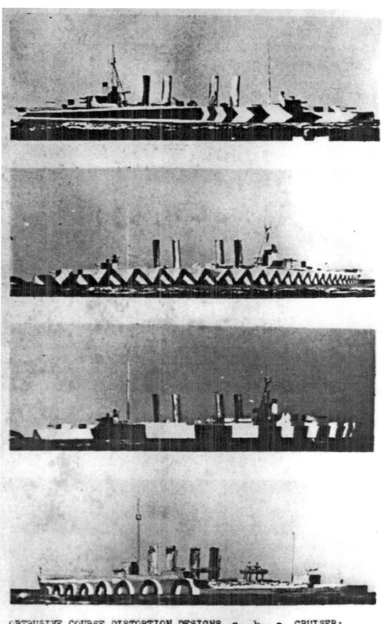

OBTRUSIVE COURSE DISTORTION DESIGNS. a , b, c, CRUISER; d, DESTROYER. BROADSIDE VIEWS.

Extreme obtrusive disruption on a series of images of a four-funnelled cruiser and destroyer contrived for the 1937 Bu-C&R publication 'Handbook on Ship Camouflage'. They show particularly well how painted patterns could create extreme perspective distortion, among them a sequence of semi-circles or a row of triangles of diminishing size along the side of the hull. In all of these images the vessels are actually broadside on. (US Navy Bu-C&R)

BRIDGE. foreside
shewing correct join of
pattern on corners.

FIG. 3

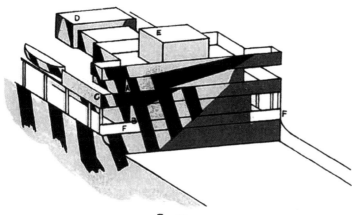

FIG. 3A

Example drawings from 'General Directions for Dazzle-Painting', Norman Wilkinson's guide to the application of dazzle paint, showing the right (Fig. 3) and wrong (Fig. 3A) methods. In this case, it refers to the decoration of the forward-facing side of the bridge, requiring the taking of the dazzle pattern around the corners of the superstructure onto the bridge front. The UK favoured this wraparound approach, whereas the USA preferred to keep the pattern on the two faces separate, believing that more confusion was caused when the corners could be seen. (Controller General of Merchant Shipping)

paintings had been exhibited pre-war. He was duly obliged with the generous provision of suitable workrooms within Burlington House, Piccadilly, London, which were occupied from June that same year.

The constitution of what is widely referred to as the Admiralty Dazzle Section is not as clear-cut as might be supposed. The team assembled for the section, some working at the Royal Academy, others based at ports around the country, where they supervised the application of dazzle designs, comprised a number holding the officer rank of lieutenant RNVR as well as civilian model makers and draughtswomen in the ratio of twenty-five of the former to twenty-four of the latter. Among the designers were many of the leading marine artists of the day, including Cecil King, who was Wilkinson's deputy, and Edward Wadsworth, who was based alternately at Liverpool and Bristol.

Similarly, whereas Dazzle painting was primarily intended for merchant ships, there were a number of warship types to which it was also applied, principally the converted liners of the 10th Cruiser Squadron, all commissioned as His Majesty's Ships, and the destroyers employed on convoy escort duties.

This close-up view of a dazzled ship's bridge and hull side clearly reveals the result of Wilkinson's instruction in practice. The contrasting bars of the pattern extending across from the hull to the bridge front make it difficult to determine the point of separation between the two structures. (Australian National Maritime Museum)

Wilkinson had evidently had to take the initiative in getting the Camouflage Section established with the Admiralty Board dragging its feet, but when first launched it came under the Directorate of Naval Equipment. Coloured dazzle plans held in Washington DC and at the Imperial War Museum, London, reveal that it subsequently transferred under the authority of the Transport Department of the Ministry of Shipping, a function created by the government under the Defence Regulations of June 1917 to regulate merchant shipping and maintain an adequate supply of wartime ships.

The Ministry was assigned control of functions previously shared between the Controller of the Navy at the Admiralty, the Board of Trade and the wartime-created Shipping Control Committee of the Cabinet. However, although responsibility for the Transport Department had been transferred to the Ministry of Shipping, it still remained under the operational administration of the Admiralty which, in 1914, had been authorised by royal proclamation to requisition merchant ships as needed for transport purposes. These organisational arrangements raised the prospect of potential discord between the military and civilian authorities.

How the activities of the Dazzle Section were actually coordinated in practice between these disparate authorities is unclear but, on the face of it, they risked compromising essential coherence of purpose, something the USA had avoided by placing all responsibility for ship camouflage within a single jurisdiction, that of the US Navy. While the Dazzle Section's staff supervised the painting of merchant ships, the painting of Royal Navy warships was presumably controlled by naval dockyard personnel, but how the designs appropriate to these different vessel types were approved and the detailed plans disseminated remains uncertain, quite besides the adjudication of orders of priority.

Central to the design efforts undertaken at the Royal Academy was the assessment of individual camouflage patterns conceived for different individual ships and groups of ship types. As part of this work, the majority of British merchant vessels were divided into thirty-seven categories, each distinguished by their fundamental differences in structural layout and length. Models of these types, produced by skilled woodworkers at the academy, were painted in different colours and patterns for viewing on a test theatre constructed by the Wilkinson team, which used a periscope apparatus for realism. Simulated sea and sky backgrounds and adjustable light and visibility facilities allowed for assessments to be conducted in conditions as close as possible to a variety of those that might be experienced operationally. An intuitive, trial-and-error approach was followed until an ideal design was ascertained.

Left: Side elevation drawing of the Royal Academy model-testing theatre. (Author, based on Norman Wilkinson's original sketch)

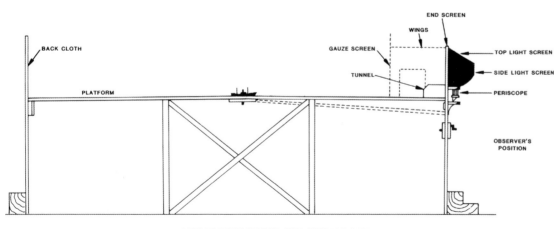

DAZZLE PAINTING SECTION MODEL TESTING THEATRE

Opposite: Norman Wilkinson working on a dazzle model at the Royal Academy. (Wilkinson estate)

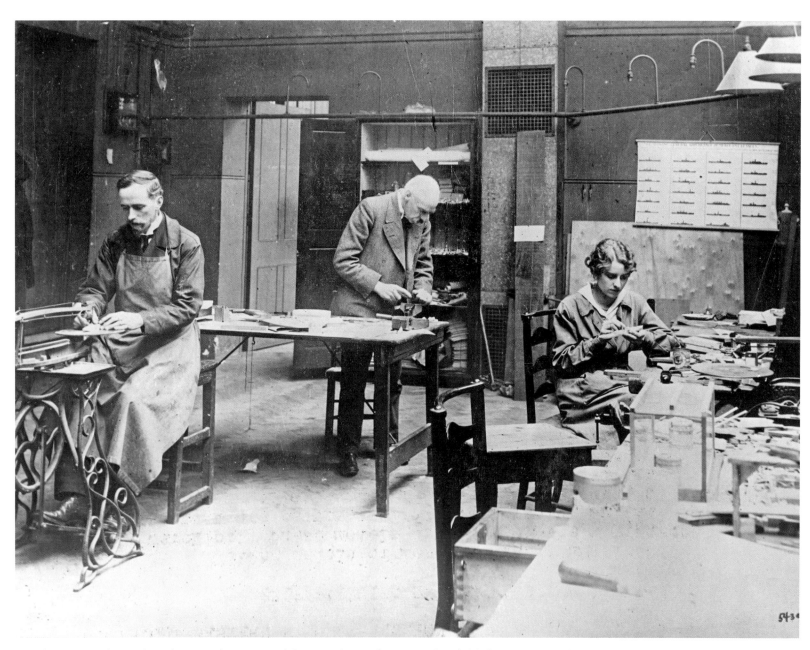

A working scene in the Royal Academy Dazzle Section workshops – making and preparing ship models for assessment in the viewing theatre. (US Naval History and Heritage Command, NH41721)

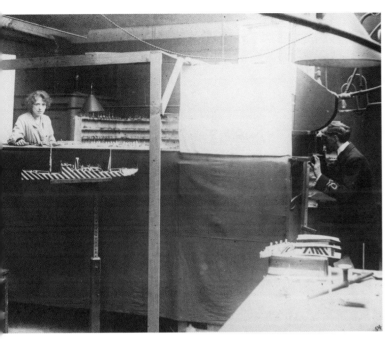

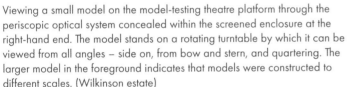

Viewing a small model on the model-testing theatre platform through the periscopic optical system concealed within the screened enclosure at the right-hand end. The model stands on a rotating turntable by which it can be viewed from all angles – side on, from bow and stern, and quartering. The larger model in the foreground indicates that models were constructed to different scales. (Wilkinson estate)

Draughtswomen at the Dazzle Section drawing and colouring dazzle plans, working from the models whose painted designs have been approved. (Wilkinson estate)

The designs and colour combinations that achieved the desired effect were then committed to plans, which were transferred to docks and ports around the country; there was a prolific number. Under the supervision of the local camouflage specialist, the shapes were implemented as accurately as possible on the ships' hulls and superstructure by various means – chalk snap lines for curves and straight boundaries, detailed shapes marked out by painters suspended on bosun's chairs or rope ladders using chalk-tipped rods, all guided by megaphone or hand signals. More painters then followed behind to fill in the spaces in the nominated colours, rather like Paint-by-Numbers. By these means, dazzle schemes could be completed in hours rather than days and, by the end of 1917, numerous vessels – troopships, cargo ships and destroyers in the main – had been sent to sea 'bedazzled'.

★★★

The process of transferring finished dazzle designs onto ships involved (a) marking out hull and superstructure areas and allocating colour codes and (b) painting the defined areas with the prescribed colours. (Roy Behrens)

The process was more demanding when larger vessels were receiving dazzle treatment, such as the White Star liner *Olympic*, shown here. The painters work off bosun's chairs suspended precariously over the ship's side. Earlier, under the direction of Dazzle Section operatives, they would have marked out the colour areas with chalk snaplines or chalk-tipped rods. (*Illustrated London News*)

The French contribution to marine camouflage development in the First World War is too often overlooked or under-acknowledged, in particular, its participation in the dazzle initiative. It was France that had given the words 'camouflage' and 'camoufleur' to the international lexicon and it was the French military authorities who, on 12 February 1915, under the direction of General Joseph Joffre, created the first ever official camouflage research and development unit at Amiens, under the leadership of Lucien-Victor Guirand de Scévola. However, its brief did not extend to marine camouflage, its primary focus being on army field camouflage.

In 1916 marine artist Pierre Gatier, a member of the unit, was posted to Rochefort, Charente-Maritime, from where, on the directive of Admiral Antoine Schwerer, he undertook an exercise to observe ships from the periscope of the submarine *Nivôse*. Whereas, sometime later, he expressed the view that 'Camouflage is more a matter of sculpture, that is, of form, than of painting', hinting at a resort to physical disguise, Gatier was among the first to promote the concept of using stark patterns and colours to confuse enemy assailants' judgement of distance, direction, speed and size. Indeed, it is reasonable to surmise that the French had arrived at the logic of the dazzle concept, if not before, then certainly in parallel with their English counterparts.

As in the UK, guidance on the protective painting of French-flag shipping at the outset of the war had been rudimentary, to say the least. Insufficient thought had been given to either the dangers confronted by merchant ships or the means by which the risk of attack might be mitigated. Essentially, the advice had been for vessels to be made less conspicuous by overpainting white areas of structure with either grey or a dull colour, eradicating painted ships' names and suppressing owners' emblematic markings on funnels.

By 1916, the French authorities had taken steps to reorganise the nation's military disposition in the light of experience of the war's evolving demands. On 10 March 1916, Vice Admiral Ferdinand De Bon was appointed Chief of the General Staff of the Navy and under his direction a Submarine Warfare Directorate (DGGSM – Direction Générale de la guerre sous-marine) was created on 18 June 1917 to take responsibility for all aspects of combating the U-boat threat.

As far as is known, it was during the three months prior to the establishment of the Submarine Warfare Directorate that tests were conducted involving some fifty ships of various size which normally operated on France's Atlantic or Channel coasts. For the trials, all the mustered ships were painted in a common dazzle-type scheme. Translated from his book *La Marine Marchande Française, 1914–1918*, author Marc Saibène explains that on their starboard side they were painted a light grey colour, over which were superimposed patches of black, light purple astern and light green amidships. On the port side, the background hull colour was light green, over which there was an irregular pattern in black with a purple patch amidships and white patches on the fo'c'sle. Funnels were painted purple to port, green to starboard. Simultaneously, a further single, auxiliary transport received a unique dazzle-type livery comprising uniform, alternating stripes in light blue, light grey and dark grey. However, having revealed indifferent painting consistency, the experiment's main recommendation was for the creation of a service to coordinate all ship camouflage practices, as much for their implementation as for their development.

Arising from a visit in June 1917 to the Burlington House operation in London by the French Naval Attaché, subsequently summarised in a report to the French Admiralty dated 28 July 1917, the decision was made to launch a French Navy camouflage section. Consideration was given to implementing a joint operation with the British but, ultimately, an independent French service was preferred which would collaborate technically with its British counterparts. Set up on 4 October 1917 (the date 21 November 1917 is also quoted), Pierre Gatier was assigned to the unit, housed in a loaned, glass-roofed studio in the Jeu de Paume Pavilion of the Musée de l'Orangerie, located in the Tuileries Gardens, Paris.

The Camouflage Section was controlled by the Service Technique des Constructions Navales (Technical Service of Naval Construction, Ministry of the Navy) as part of the Submarine Warfare Directorate. It was headed by First Lieutenant Henri de Lamothe-Dreuzy with, as his deputy, Second Lieutenant Eugène Ronsin. Besides Gatier, the two other designers on the unit's strength were Georges Taboureau and Léon Félix. Three model-makers and five draughtswomen, who produced the final plans, completed the complement. As such, with a

Staff of the French Naval camouflage centre in the Jeu de Paume Pavilion of the Musée de l'Orangerie, Paris, on 21 November 1917, with Pierre Gatier central at the back, wearing a cap. The top-secret operation occupied the first floor of the pavilion. As in London, following assessment, the approved French designs, each complete with its allocated colour palette, were drawn up on fine art paper for release to the designated port where application was scheduled to take place. (Roy Behrens, photograph attributed to P. Dantec)

EXIGER DANS L'EXÉCUTION DE CES PLANS QUE LE **TRACÉ** DE CAMOUFLAGE SOIT FAIT **AVEC LE PLUS GRAND SOIN.**

LES NUANCES DOIVENT ÊTRE RACCORDÉES **EXACTEMENT** AVEC LE CARNET D'ÉCHANTILLONS DE COULEURS.

LES COULEURS EMPLOYÉES DOIVENT ÊTRE SUFFISAMMENT COMPACTES POUR BIEN **COUVRIR** LES SURFACES À PEINDRE. AUCUNE COULEUR NE DOIT ÊTRE **BRILLANTE.**

AUCUNE PARTIE DES NAVIRES, APPARENTE DE L'EXTÉRIEUR, NE DOIT, SOUS AUCUN PRÉTEXTE, RESTER NON CAMOUFLÉE : LES PARTIES EN BOIS VERNI NE SONT PAS EXCEPTÉES DE CETTE RÈGLE.

LE CAMOUFLAGE DOIT TENIR COMPTE DANS SON EXÉCUTION DE LA PLACE QU'OCCUPENT **EN RÉALITÉ** LES EMBARCATIONS À LA MER.

A directive on camouflage practice issued by the Marine Nationale in the First World War. The instructions call for the execution of plans with care, in strict accordance with the supplied designs which have been prepared taking into account ships' zones of operation. Exact shades as recorded in a colour sample book must be used with sufficient well-coverage of all surfaces including varnished wooden parts. No surface was to remain uncamouflaged. (Service Historique de la Marine Nationale, courtesy of Jean-Yves Brouard)

total of thirteen personnel, the French section was barely a fifth of the size of its British equivalent, although it must be remembered that the British merchant fleet was then more than nine times larger than that of the French.

Along with two others, who would become members of the French team, Gatier also paid a visit to Burlington House, London, in September 1917 to view the equipment there and observe how the British unit went about evaluating their designs. Arising from that mission, a viewing theatre arrangement comparable to that erected at the Royal Academy was constructed in the Jeu de Paume workshop, equipped with a trench periscope.

The experience of Eugène Ronsin as a pre-war painter of stage scenery proved invaluable for the creation of realistic marine backdrops against which test models could be observed. Each of the 20cm-long test models was based on 1:200 scale side elevation drawings which all merchant ship owners were directed to submit for evaluation, along with descriptions of characteristic structural features and details of the sea areas in which their vessels routinely operated.

Implementation of approved French schemes at commercial ports was the responsibility of a naval officer commanding the local military authority (AMBC – L'armament militaire des bâtiments de commerce), who engaged civilian contractors to undertake the work. Within naval dockyards, the painting of warships was executed by the naval construction service. The French camouflage section continued its work through to the armistice on 11 November 1918, when it was disbanded.

★★★

Opposite: Artist's impression of the dazzled storeship *Industry*. (Lieutenant Jan Gordon RNVR)

Just as the US camouflage schemes devised by Mackay, Toch, Herzog and others gradually evolved, so too did British and French dazzle. It became apparent that it was not necessary to use gaudy primary colours to achieve the desired perceptual effects – they were attained equally well using just black, white and shades of grey and blue.

The use of white paint remained controversial. Despite Thayer's insistence that it constituted the best colour for ship camouflage in all conditions, complaints emanating from fleet units expressed concern at its greater conspicuity in certain conditions of light. Thus, under an Admiralty directive, mixtures of light grey, grey-blue or grey-green were substituted for white, a change of practice that was communicated to the US Navy Department on 23 May 1918.

The debate over the benefits, or otherwise, of white, like the enduring disagreement between the opposing advocates of low visibility versus disruptive painting, was to rumble on through to the Second World War. In the short term, this issue was to have collateral ramifications, in part prompting an Admiralty review of dazzle as a whole, which led on to a legal dispute as to its provenance.

Pattern change was another mutation in the British Dazzle portfolio towards the end of the First World War, with a preference for a wasp-like, zebra-striped design variant that accentuated structural disembodiment and false perspective. In the USA, from January 1918, dazzle painting was adopted to the exclusion of all other ship camouflage systems, yet the processes of experimentation and development, as well as the resulting styles, followed along somewhat different lines to those in the UK and France.

The following photographs show examples of a variety of British or Admiralty Dazzle designs on different ship types. The designs exhibit many of the artistic components and motifs that were employed, although the precise intention in each case is not necessarily discernible.

The dazzle design of Cunard's *Mauretania* incorporated distinct texture-gradient elements on the forward section of the hull. (US National Archives 165-WW-274A-024)

Commonwealth Government Transport Ship *Bulla* (formerly *Hessen*) exhibits a dazzle pattern of irregular polygons and curves. (A.C. Green)

Built by Swan Hunter and Wigham Richardson on the Tyne, *War Climax* is seen at completion on 28 September 1918. The employment of bold, contrasting, angular lines causes significant shape confusion, suggesting a double bow. (Tyne & Wear Museums)

An unidentified cargo ship, possibly French, alongside at the La Ciotat Shipyard, Occitan and Provençal, France. (Jean-Yves Brouard collection)

Left: Bibby Line's *Leicestershire* has a similar dazzle design to that of *Bulla*. (A.C. Green). Right: The destroyer HMS *Talisman* has a design of angled, almost vertical stripes in three or four shades of colour. The spacing of the stripes and their increasing width was probably intended to distort perspective to induce confusion of bearing. Larger naval vessels, including battleships, battlecruisers and the aircraft carrier *Furious*, were also dazzle painted. (Adrian Vicary)

Left: The fate of the troopship *Justicia* (formerly *Statendam*), torpedoed and sunk on 20 July 1918, perhaps raised questions about the credibility of dazzle painting. Realistically, though, assailed from either quarter in a combined attack by two U-boats, no camouflage system could have protected her in such a situation. (US Naval History and Heritage Command, NH101616)

Opposite: Union-Castle's *Walmer Castle*, as taken up for war duties, wears a style of dazzle that was adopted for many of the British troopships. (Tom Rayner)

Another example of this dazzle design is seen on the Orient Line's *Osterley*. Other ships to wear the scheme were *Empress of Russia* and even, briefly, White Star Line's *Olympic*. (Richard de Kerbrech collection)

The aircraft carrier HMS *Argus* has a similar dazzle design, which is not surprising since her construction had begun as the Italian passenger ship *Conte Rosso*. Her deep-hulled liner form lent itself to the same treatment. (US Naval History and Heritage Command, NH63225)

Towards the end of the war, an intense form of striped dazzle was favoured by the UK using blues, greys, black and white, as shown here on the Huddart Parker steamship *Zealandia*. (Richard de Kerbrech collection)

Pacific Steam Navigation's *Orissa* has a similarly intensive, striped pattern all over but, in her case, with the addition of a triangular device on her bow. (A.C. Green)

The dazzle pattern of USS *Leviathan* (formerly *Vaterland*) was designed at the Royal Academy workshops by a member of Norman Wilkinson's team. The sawtooth arrangement along her portside flank and bow was a particular artifice designed to distort perspective and undermine judgement of bearing. It was also utilised on Cunard's *Aquitania*. *Leviathan* is seen alongside at Pier 4, Hoboken, in 1918. (US National Archives, 165-WW-274A-022)

A dazzled French Line (CGT) liner taken up for troop transportation, the 1913-built *Puerto Rico*. (Marc Saibène collection)

From the autumn of 1917, as the French camouflage section commenced output of disruptive designs, French ships such as CGT's *La Lorraine*, seen here, began to appear with dazzled livery. (Jean-Yves Brouard collection)

Thirty-six American destroyers of Admiral William Sim's Queenstown Squadron had dazzle designs prepared for them at the Royal Academy Dazzle Section, London. USS *Ward* off Mare Island, California, in September 1918. (US Naval History and Heritage Command, NH50265)

USS *Kimberley*, also of the Queenstown Squadron, photographed off the Irish coast. (US Naval History and Heritage Command, NH51080)

Visual Tricks 1

DISAPPEARANCE ILLUSIONS

The abstract designs, patterns and colours, in all their variety, that were devised to deliver disruptive camouflage protection incorporated additional visual tricks intended to aid in the deception of U-boat commanders and cause the assailant to discharge his torpedoes wastefully. Whether intentionally or by chance, the most effective of these manipulated a perceptual phenomenon derived from the principles of the Gestalt Theory.

Through the application of paint colours in pronounced, contrasting tones to adjacent structural features or sections of hull, effectively 'painting in' and 'painting out' to deliberately undermine cognitive unit-forming, it was possible to make certain of these systemic parts 'disappear'. This was not about the painting of a single constituent feature in, for example, bands of alternating colours, but entire, separate structural components, relatively close to one another, in either a dark tone or a light shade. By so doing, recognition of vessels for what they were or how they were assembled was made difficult. More importantly, it undermined comprehension of their dynamic behaviour.

The manifestations of the painting-in-and-out artifice as a means of inducing sensory confusion when observing ships were essentially twofold: first, hull foreshortening by painting in white paint or an extremely light grey the complete bow or stern section of the hull (or often both), while the remainder was painted in a dark tone (occasionally the complete reverse); second, where ships such as troop transports had multiple funnels, darkening and lightening them alternately, or in pairs. A four-funnel ship might thus appear to have only two funnels.

Using white or very light grey in this way, as the 'hiding' pigment, hinted at lending credence to Thayer's claims regarding these tones as being the ideal general concealment colours. But that would be an oversimplification. It is the interaction between the adjacent contrasting tones that caused the illusion of disappearance not, necessarily, the use of white or light paint alone. As Lieutenant Jan Gordon RNVR, an artist attached to the Dazzle Section in London, had to say, 'Disruption does to a certain extent combine with an amount of low visibility, but it must be insisted that the chase [pursuit] of actual invisibility is a myth which will never be solved.'

The selective application of white or a light shade to exploit the particular visual phenomenon of 'hiding' recognisable structural features was certainly valid, but the widespread use of such colours for low visibility remained questionable. As the following pictures demonstrate, the practice of 'painting in and out' continued to be utilised extensively in both world wars.

Former Norddeutscher Lloyd liner *Kronprinzessin Cecilie*, as the American troop transport USS *Mount Vernon*, seen berthed at New York on 8 July 1918. Her forward two funnels have been 'painted out' with light-coloured paint, while the aft two have black stripes down them, to give the impression in misty conditions or low light levels that she was a twin-funnelled ship. (US Naval History and Heritage Command, NH45747)

Above: Another American troopship given the painting-in/out treatment was USS *Louisville*, the former American Line ship *St Louis*, here also seen at New York. Her forward funnel has all but disappeared in the smoke and overcast sky. (US Naval History and Heritage Command, NH51431)

Left: In similar fashion, USS *Von Steuben* (formerly Norddeutscher Lloyd *Kronprinz Wilhelm*) which, apart from a black vertical stripe on her second funnel, has had all her remaining funnels 'painted out' in order to disguise her. The light-coloured forward hull overpainted with contrasting angular radiating stripes was a typical device for distorting perspective to confuse U-boat commanders. (US Naval History and Heritage Command, NH42418)

Clockwise from above: The effects of selective painting out are best appreciated when ships are seen underway against a sea background, such as in this example showing USS *Northern Pacific* of the Great Northern Pacific Steamship Co. Parts of her recognisable hull and superstructure features are lost in the background. The painting-out process was found to be less effective in bright, clear atmospheric conditions. (US Naval History and Heritage Command, NH78285); Making use of white or light paint as a partial concealment colour appears, on the face of it, to vindicate Thayer's low-visibility doctrine for regions beset with overcast climatic conditions. In some respects, the process of painting in and out would appear to support the concept of employing disruptive colouring as a background-blending mechanism rather than for size, identity or distance confusion. However, in certain circumstances, white or light-coloured paint across an entire ship or applied in large, unbroken areas could accentuate its conspicuity. Indeed, reinforcing this dichotomy, under the Geneva Convention, hospital ships were required to be painted all white expressly to make them more conspicuous. This is USS *Mongolia*. (US Naval History and Heritage Command, NH105722); In this view of USS *Leviathan*, seen from an escorting destroyer, the white areas of her paint scheme have disappeared into the murky background, the eye drawn to the darker colours, such that she is not seen as a complete entity. (US Naval History and Heritage Command, NH101625)

In the Second World War, Liberty ships and repair ships like USS *Mindanao*, shown here in Hampton Roads on 23 November 1943, had their funnels and mast tops whited out, effectively to hide them against the horizon sky. Whereas her funnel cap has been omitted from this attention, the upper ends of the derrick arms and the gun emplacements have also been painted out. (US National Archives)

The French Chacal-class destroyer *Jaguar* has her entire forward end painted white. Further aft she has a curious, white-painted semicircle on the dark section of her hull. (Jean-Yves Brouard collection)

Clockwise from top left: A photograph taken by Kazutoshi Hando of the Japanese warship HIJNS *Kiso*, said to date from 1924 but probably taken in the Second World War. She has a quite complex camouflage scheme combining a pattern of striated lines in contrasting colours with whited-out bow section. (US Naval History and Heritage Command); Whiting, or painting out, has given the German battleship *Tirpitz*, seen here moored in a Norwegian fjord, a false bow and stern, a foreshortening device used to suggest ships were at greater range. Besides that, her camouflage scheme exhibits additional disruptive features. (US Naval History and Heritage Command, NH71390); The complex dazzle design of *Andrea Doria*, photographed at Trieste, reveals a whited-out false bow and stern along with fake bow waves. (Ufficio Storica)

Sister ship of *Tirpitz*, *Bismarck* can be seen docked during painting by her crew. The forward section of her hull has been painted out in a darker shade to create a false bow on which a rudimentary fake wave has been added to amplify the deception. (Mary Evans Picture Library)

The Italian cruiser *Luigi di Savoia Duca Degli Abruzzi*, photographed when surrendering at Malta with other naval units under the 1943 Armistice, reveals how effective the practice of painting out of the bow area could be. (Crown Copyright)

Like *Bismarck*, the heavy cruiser *Prinz Eugen* has been foreshortened using dark paint at the bow and stern rather than a light tone. Although the midships hull has not been painted white, it has a sufficiently lighter tone to achieve the same effect. Chevrons in contrasting colours add to her disruptive scheme. (Jean-Yves Brouard collection)

4

UNHITTABLE RATHER THAN INVISIBLE – US DISRUPTION INITIATIVES

As stated earlier, disruption painting took multiple forms, each with different objectives in mind. 'Dazzle' was the term coined for Norman Wilkinson's extreme variant but it was certainly the case that not all dazzle was the same. In simple terms, this can be expressed as follows: while all dazzle is disruption, not all disruption is dazzle and not all dazzle is the same dazzle. Indeed, it was claimed that it was possible, by looking at different disruptive designs, to determine their country of origin, although, it has to be said, it is not that easy in the absence of original, annotated plans to differentiate between them in black and white photographs when the colours appear only as tones of grey.

Although Abbott Handerson Thayer had first proposed the concept of disruption as a means of breaking up structure to facilitate concealment by background blending, it was Everett Longley Warner who was first, in the USA, to promote the use of disruption to interfere with perception of target range and course. And, like Norman Wilkinson, his scheme of obtrusive distortion was intended for merchant vessels rather than warships.

Everett Warner, who hailed from Vinton, Iowa, was an established artist, known for his impressionist paintings and prints. His disruption dazzle scheme, approved on 1 October 1917, formed a key stage in the evolution of the particular form of disruption dazzle ultimately adopted by the US Naval authorities.

It can be conjectured that Warner had been actively developing his unique dazzle scheme prior to when he was diverted, in September 1917, to assist Thomas Edison in an attempt at dynamic disguise in which either the donated Cunard ship *Valeria* or the captured German freighter SS *Ockenfels* (reports conflict as to which it was) was clad in a wood and canvas framework to totally conceal its structural features. The experiment was an unqualified failure, largely because of the unseaworthy nature of the shroud when exposed to conditions of turbulent wind and weather – that is, with the exception of Warner's contribution to the project, the painting of the hull in an abstract camouflage pattern.

No doubt as a result of the failure of that endeavour, Warner concentrated his attention on disruptive devices as an alternative, more promising stratagem. Contrasting his dazzle scheme with Wilkinson's correlative system, it can be seen that, in essence, they were quite different. Less geometric with few, if any, straight lines, Warner's was characterised by bold, sweeping, abstract shapes and stripes painted in pronounced, contrasting shades of red, blue or green with white or, in monochromatic versions, in grey tones along with black and white, the emphasis in all cases being on elevated contrast.

Warner placed great stress on the value of intense contrast to cause confusion, using white paint as an outline between the other coloured or toned areas to amplify the effect, because he considered this to be the best way to create a deceptive appearance. The intention was to dismember the recognisable lines and contours of a ship at short range. However, the cause of reduced visibility had not been entirely abandoned since his selection of primary colours when distributed in properly balanced ratios would, he believed, blend into an overall pale grey at greater distances. But his declared principal objective remained making a vessel 'hard to hit rather than hard to see'.

Warner's scheme was not only implemented for some larger troop transports but also to certain naval ships. Nonetheless, it was a short-lived practice for, along with all the other US schemes originally approved, it was abandoned when, in January 1918, the Navy Department announced that it was to adopt, without exception, dazzle painting of the Wilkinson type. It was a declaration which, along with the imminent creation of the US Navy's Camouflage Section, was to irrefutably clarify the US position regarding matters of ship camouflage.

It is apparent that, despite the navy's emphatic pronouncement, the Submarine Defense Association still saw the new section as the vehicle by which its own schemes, developed by the Loyd Jones team at Eastman Kodak, could gain official sanction. In the Submarine Defense Association Report of 8 March 1918, Lindell T. Bates declared, 'It is anticipated that the governmental authorities will direct that vessels conform to the recommendation reached by the Association's staff and experts prior to the creation of the naval camouflage unit, of which its two aides have become the leaders.'

The two aides to whom Bates was referring were Loyd A. Jones and Harold Van Buskirk. In fact, Van Buskirk was appointed as the overall head of the entire Navy Camouflage Section, while Loyd Jones was to lead the Research Sub-Section at Eastman Kodak, Rochester. Engaged to head the Camouflage Design Sub-Section in Washington was Everett Longley Warner. All three men were commissioned as navy lieutenants for the duration. The Camouflage Section, as part of the Maintenance Division, ultimately reported to David W. Taylor, Chief of the Bureau of Construction and Repair and Chief Constructor of the Navy.

Secretary of the Navy Josephus Daniels wasted no time in unequivocally spelling out the navy's camouflage agenda which, at a stroke, quashed any remaining aspirations that the Submarine Defense Association still held. In his letter of 25 March 1918, sent to Henry C. Grover, in charge of the US Shipping Board's Emergency Fleet Corporation, Daniels outlined the co-organisational arrangement that had come into effect, making it abundantly clear that all future camouflaging of Shipping Board ships was to be prescribed and supervised by the Navy Department. His exact words were:

(a) The Navy Department (Bureau of Construction and Repair) will have charge of all camouflage designs, which will include the investigation of suggestions as to schemes of camouflage and the issuance of definite instructions as to the type of camouflage to be adopted.

(b) The Emergency Fleet Corporation (The Shipping Board) will have charge of the practical application of camouflage to vessels.

(c) The design furnished by the Navy Department (Bu. C&R) is not to be departed from except where it may be necessary (owing to discrepancy between the dimensions shown on the plan and those of the actual ship) to expand or contract a portion of it. The color[s] is not to be departed from. No actual redesigning is to be done nor are colors not in the plan to be introduced.

Such an unambiguous stipulation of the modus operandi to be adhered to, along with the declaration that, without specifying its precise form, dazzle was to be adopted for all US flag vessels, tended to compound the rift between the opposing champions of low visibility versus disruption. It is not clear how Loyd Jones felt about these decisions but it must have been a heavy cross for this staunch advocate of low visibility to bear, finding himself the head of a sub-section of an agency now pursuing a policy in direct conflict with his beliefs, which

Lieutenant Everett Warner, in charge of the Camouflage Design Sub-Section at Washington DC, seen examining a model of a four-funnelled troopship on the section's viewing range. (US National Archives, 165-WW-70)

he considered he had precisely validated by scientific observations and unique mathematical algorithms.

The records do not tell us much about the ongoing relationship between the two sub-branches of the US Navy Camouflage Section but it is thought that, from that point, Jones and his team concentrated their efforts on studies of visibility and colour behaviour from which other measurement devices emerged. After the war's end, Jones engaged in sea expeditions during December 1918 to investigate weather coefficients using the colorimeter he had invented, which was subsequently patented on 3 June 1924 (Patent No. 1,496,374). In February 1919, its commission at an end, the Rochester Camouflage Sub-Section was closed down.

Whether or not William Andrew Mackay was offered a position with the Navy Camouflage Section is unknown but, either by his preference or because he had been overlooked, he instead joined the New York region camouflage unit of the Shipping Board's Second District. Divided into districts encompassing the entire coast of the USA, each with a senior camouflage officer, the Emergency Fleet Corporation's strength of trained camoufleurs rapidly numbered in the hundreds.

In short order, Mackay had his camouflage training school equipped with its own independent model-viewing theatre. All the indications are that Mackay and Warner did not see eye to eye, an understatement perhaps, and it is clear that Mackay and other Shipping Board camoufleurs were unhappy with the new organisational arrangements. It seems there was resentment at the US Navy's imposition of having its own artists promulgate all the ship camouflage plans, both military and mercantile, while leaving the Shipping Board operatives to do all the dirty work. The Shipping Board teams considered themselves perfectly capable of devising, testing and approving designs, especially as the majority were destined for merchant vessels. John D. Whiting, who worked with Mackay in New York, shed some light on this in his book, *Convoy: A Story of War at Sea*, quoting one of his disgruntled colleagues:

I thought these designs were made in Washington, under the Navy Department. Yep, that's the theory. Those Johnnies think we can only put their patterns on the steamers but bless you, the plans we get from there only fit half the shapes in this town.

He [Mackay] is quietly evolving his own school for camoufleurs [discrete from the navy].

Ironically, it was Mackay, rather than Warner, who was to feature prominently in the US Navy's continuing camouflage protection endeavours throughout the inter-war period, activities which culminated in the first comprehensive guide to ship camouflage practices published by the Bureau of Construction and Repair on 15 September 1937 under the title, 'Handbook on Ship Camouflage'. It was largely the product of Mackay's labours, his contribution to US Navy camouflage research and development only brought to an abrupt end by his untimely death through a heart attack on 26 July 1939.

In order to distinguish its type of disruptive painting from British Dazzle and because the adjective 'dazzle' was considered to be misleading and insufficiently explicit as to the system's purpose, terms like 'baffle painting' and 'jazz painting' were coined instead in the USA. In the event, though, neither of these alternative idioms caught on.

For the remainder of the war, dazzle tended to be differentiated as being either 'British' or 'US Navy'. And there was a difference between them, largely emanating from the distinctive approach to design evolution and evaluation adopted by the Everett Warner team. Whereas the method applied by the Royal Academy operation had drawn intuitively on relatively limited artistic and geometric principles, followed by a trial-and-error process of refinement, the US procedure was to become more analytical, arising from which dissimilar designs emerged.

Despite the differences, there was a fair degree of transatlantic collaboration, in the sharing of concepts and the need, through service feedback, to make adjustments in the light of operational experience. Arising from a visit to the Royal Academy by Vice Admiral William Sims, who was in command of all US Naval forces operating from the UK, Rear Admiral Clement Greatorex, the Director of Naval Equipment at the Admiralty, requested Norman Wilkinson to address a letter to the US Navy Board with plans for three typical dazzle designs for cargo vessels. In his covering letter, by way of guidance to the US Camouflage Section, Wilkinson explained:

The pattern must not be too small or the result will be recognizable at a short distance, nor too big as in the latter some large proportions of the vessel will be definitely shown.

The size, length and bulk of the vessel to be painted must be considered.

Only colours easily procurable [to be] used.

And, to close, a comment of considerable significance: 'The use of lights and shadows in accordance with Thayer's Law is incorporated where necessary.'

Wilkinson's letter was followed up, on the invitation of Vice Admiral Sims, by a personal visit by him to the USA in March 1918 to provide further assistance. Armed with ninety more plans, Wilkinson sailed aboard USS *Leviathan*, the former German liner *Vaterland*, a ship for which his team was to create a unique dazzle design.

As Everett Warner explained later, for all this facilitation, it became apparent that there was an absence of a defined underlying axiom by which the principles of dazzle, as developed in the UK, could be consistently applied. Although, like his British counterparts, he too had initially followed a trial-and-error technique, he was convinced that there had to be fundamental rules to the dazzle illusory process which, if found, could assist rapid development of effective designs.

Among Warner's team were many of the leading US marine artists of that time, including Frederick Judd Waugh, who was invited to join the operation at the mature age of 57 years. It was an advantageous appointment because Waugh was a most capable designer with an intuitive feel for the practice of perspective distortion, notably the mechanism of reverse perspective, where the stern of an approaching ship was made to seem nearer than its bow. He developed some of the most striking and highly effective US dazzle patterns as exhibited on, among others, such ships as the NOTS (Naval Overseas Transport Service) vessel *West Mahomet* and the naval oiler USS *Proteus*. In every respect, he endorsed the core principle of Warner's fundamental philosophy: to alter rather than to conceal.

The outline technique initially adopted by the Washington organisation, more or less comparable to that followed in London,

Naval ensigns making wooden models in the workshop at the Washington DC Camouflage Design Sub-Section. (Roy Behrens)

comprised a logical sequence: first, make scale models of each type, identifying inherent structural peculiarities and apply a tentative pattern accordingly; then view the models on the specially constructed testing theatre at a representative distance of 2,700 yards; make alterations, since, more often than not, they would be required before an acceptable design was reached for approval; transfer the finished, painted models to the draughting department for the designs to be transposed by visual interpretation into two-dimensional form on special-type plans; finally, dispatch the signed-off, finished plans to the camouflage districts for application.

Feedback on the adopted designs, provided by observers from the Office of Naval Intelligence, permitted the Camouflage Section to refine its creations. Using a bespoke reporting method, they supplied photographs, sketches and written reports covering the appearance of the colours, the nature of the weather and any other factors witnessed that had a bearing on the efficacy of the designs.

Returning to Warner's desire to augment the potency of the dazzle illusions, a way forward was found quite by chance when an exercise was carried out to analyse the team's earliest designs. It was discovered that those which were the most effective had resulted, unintentionally, from either the arbitrary use of particular geometric forms or where the ships' structural idiosyncrasies had been evaluated more critically.

Certain shapes and movements of line had caused unusual visual effects to occur. Through further investigation to understand the causes, it became apparent that it was because, unlike a two-dimensional projection painted onto a flat two-dimensional surface, these designs appeared to have been contrived in such a way that, when painted on the surface, they behaved as a geometric solid, especially when viewed from particular positions.

Up to this point, dazzle patterns had induced the effect of 2D perspective as a dimension of plane geometry by the use of shapes such as angled lines, curves and rectangular outlines painted on a 2D flat surface – 2D images giving a 3D illusion. Warner's team began to explore the creation of distortions based on solid geometry, whereby shapes like cubes, cylinders and stepped forms would be painted onto a flat surface to cause 3D optical illusions – 3D shapes amplifying the 3D illusion. Everett Warner later stated that he considered the discovery of

Another scene at the Washington DC Camouflage Design Sub-Section. The models are painted by more of the team's camoufleurs. Everett Warner is far left. To the right of him, with a beard, is Frederick Judd Waugh. The slogan 'keep it simple' was particularly pertinent to disruptive painting. Some designs combined complex elements, resulting in greater time needed to apply and maintain them. (Roy Behrens)

After assessment on the viewing range, the final stage of the dazzle disruption process was transferring the designs to plans, as seen here. Everett Warner checks the work of one of his colleagues. (Roy Behrens)

One of the female camoufleurs at the Washington DC Camouflage Design Sub-Section engaged in scale-model preparation, fitting the lattice masts of a model battleship. For realism, much attention was given to detail in the making of test models. (Roy Behrens)

this geometric phenomenon and its subsequent utilisation as the USA's chief contribution to the evolution of dazzle camouflage in the First World War.

Besides the practical implementation of these effects, novel procedures were also first experimented with and then adopted for routine day-to-day design development. One of these involved the preparation of two duplicate models of the same vessel, the first left unpainted and the other painted in a test dazzle pattern. The models were then placed on the viewing range at right angles to one another and studied at eye level. The painted pattern, as seen in perspective in one plane, was then transferred onto the unpainted model in the opposite plane, creating the illusion that it appeared to be facing in one direction when, in fact, it was facing in the complete opposite.

Another procedure, again involving the use of a pair of identical models, called for the dissection of the painted version into segments. These segments were then arranged in curves or angles, such that they influenced the way the perspective appeared to the eye, accentuated in either one direction or both. The resulting distortion effect when applied to the unpainted model gave the impression that it was headed away acutely in either a forward or aft direction, or that it was structurally bent in a curve from amidships.

In both these and other cases, pronounced visual distortion of true bearing could be caused but, in addition, when vessels painted in this fashion were viewed from either the bow or stern quarter, the hull side no longer appeared to be flat and perpendicular but often gave the impression of protruding outwards in steps. Quite apart from the visual interference already experienced by would-be attackers as they endeavoured to obtain measurements to take aim, these illusions introduced a whole additional level of disturbance to their perception.

Along with these revolutionary advances, the Washington Camouflage Design Sub-Section, like its British counterpart, also came to appreciate that vivid, primary colours – which could be easily neutralised by filters added to the periscope systems of enemy submarines – were not essential to achieve the desired results. Thus, the colour assortment in use was reduced down to tones of dark, medium and light grey, typically tinted with green or blue, plus pure blue and green shades and, where appropriate, black and white.

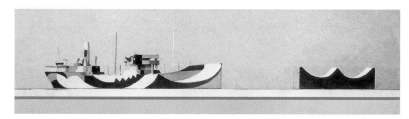

An illustration of one of the techniques introduced by the Washington DC Camouflage Design Sub-Section for creating disruptive designs to influence bearing perception. In the top view, the model appears to be bearing away to port. The bottom view shows that, in fact, it is heading to starboard, towards the viewer. (Roy Behrens)

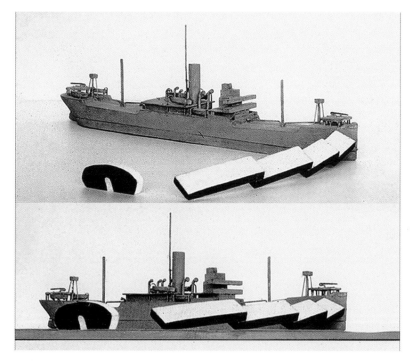

To demonstrate the simplicity of his concept, Warner produced this example by randomly arranging overlaid wooden blocks and transferring the design they visually created onto the model hull beyond. (Roy Behrens)

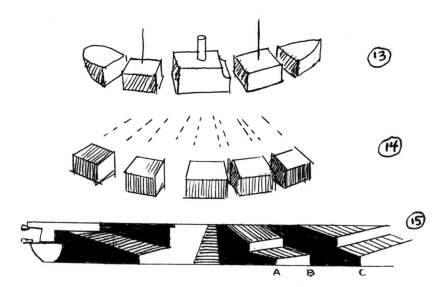

Left: Warner's drawing, taken from his unpublished 1944 guide, illustrates the principle of using dissection to produce solid, geometric disruptive designs. The side elevation at the bottom shows a solid, geometric dazzle design for a Second World War escort carrier. (Roy Behrens)

Below: Models demonstrating the Warner solid geometry concept, here of Type 9, Design K. The design was evolved from dissecting the model on the right, which also has its bridge and accommodation structure offset to port. (Roy Behrens)

Bottom: Another view illustrating the effect of the Type 9, Design K disruptive scheme, showing models of a ship with and without dazzle paint. (Roy Behrens)

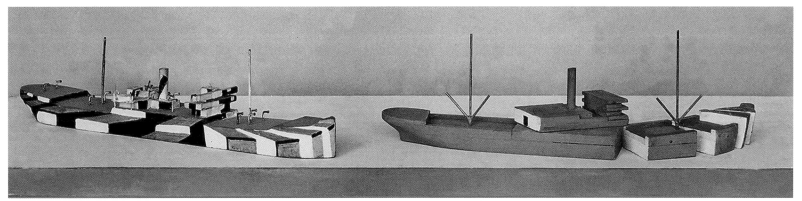

Naval Overseas Transportation Service (NOTS) SS *Isanti* is a good example of the Type 9, Design K scheme in practice. It is so effective that the hull side has the illusion of being stepped outward in boxlike shapes. (US National Archives, courtesy Roy Behrens)

In total, the Camouflage Design Sub-Section produced 495 different camouflage designs, 302 for application on merchant ships, the remainder for use on warships. These contributions, along with those of its Allied camouflage counterparts, together with the convoy system and anti-submarine patrols, had a profound impact on turning about the Allies' fortunes in the U-boat war compared to earlier periods (see Chapter 3). The shipping losses of 1918 showed a reduction of 56 per cent on the previous year, decreasing almost to 1916 levels.

The end of the First World War on 11 November 1918 led to the winding down of much ship camouflage effort, including the Warner team's new concepts still, then, in their infancy. Although this prevented their application to the extent desired, like other measures pioneered in that war, they found their way into the camoufleur's toolbox and some of them, at least, resurfaced twenty years later.

	First Qtr	**Second Qtr**	**Third Qtr**	**Fourth Qtr**	**Total**
1918	581	494	468	104	1,647

Opposite: Edison's attempt at dynamic camouflage on a cargo ship. The enclosure of the ship's structure failed in bad weather but note the hull pattern designed by Everett Longley Warner. It is uncertain whether the ship in the picture is *Valeria* or *Ockenfels* since their general silhouettes and the lifeboat and derrick positions of both vessels were virtually identical. (Roy Behrens)

Some examples of Warner Dazzle, introduced prior to the wholesale adoption of US Navy Dazzle. As with British Dazzle, the precise objectives in many cases are indeterminate and cannot readily be deduced.

The troop transport USS *Louisville*, formerly *St Louis*, in a classic Warner scheme, seen prior to being painted in a dazzle design. (US National Archives, 19-N-7573)

This view of the former Ward liner *Siboney* shows another good example of Warner Dazzle. The extensive use of white paint is striking, given the controversy surrounding its use. (US National Archives)

Siboney's sister ship *Orizaba* also displays a striking Warner Dazzle design. During the Second World War she was painted in the US Navy's Measure 32, Design 11F dazzle. (US National Archives, 19-N-12185)

USS *Plattsburg*, originally Inman Line's *City of New York*, in a photograph taken at the New York Navy Yard. Her camouflage suggests she is wearing a Warner Dazzle design but as the photograph is dated 7 June 1918 it is actually US Navy Dazzle. (US Naval History and Heritage Command, NH43047)

Another Warner lookalike, the design of the British troopship *Balmoral Castle*. Generated by Norman Wilkinson's Royal Academy operation, it featured attributes such as the irregular, outlined curves, very similar to those exhibited in Warner Dazzle. (Tom Rayner)

The destroyer USS *Caldwell* also appears to be in a Warner Dazzle design. The Warner concept was only applied over a short period before it was overtaken by the US Navy's blanket decision to universally adopt a practice more like Admiralty or British Dazzle. It is not known how widespread the use of Warner Dazzle was, or whether it was in fact applied to naval vessels at all. (US Naval History and Heritage Command, NH52854)

Next, examples of the many forms of US Navy Dazzle as applied to troop transports, cargo vessels and NOTS ships, oilers and warships.

The dogtooth triangular shapes along the waterline of USS *Pocahontas* (formerly *Prinzess Irene* of Norddeutscher Lloyd) may have been intended to break up her overall shape as well as to distort perspective. Her bow area has been painted in light shades like those recommended by the Submarine Defense Association. (US Naval History and Heritage Command, NH68721)

USS *Lenape* at New York on 20 August 1918. In peacetime, she operated under the same name for the Clyde Line. (US Naval History and Heritage Command, NH51274)

USS *Harrisburg*, formerly *Philadelphia* and *Paris* of American Line, seen in the same drydock at the New York Navy Yard. Her design is almost identical to that of *Plattsburg*. (US National Archives, 19-N-1479)

The oiler *John M. Connelly* at Philadelphia on 6 May 1918. (US Naval History and Heritage Command, NH65096)

O.B. Jennings. The long, low hulls of oilers and tankers presented particular challenges for camouflage designers. (US Naval History and Heritage Command, NH44574)

The destroyer USS *Trippe* alongside at Queenstown (now Cobh), Ireland, has dark colours concentrated centrally on her hull, a typical foreshortening trick. (US National Archives, 165-WW-335E-038)

USS *Schley*, seen at the US Submarine Base, San Pedro, California. (US Naval History and Heritage Command, NH78613)

Panaman also has a foreshortening design but, in her case, with a disconnected dark area at the bow. (US National Archives, 19-N-1114)

An example of a disruptive design feature intended to contrive reverse perspective on the cargo ship *Santa Olivia*. (US Naval History and Heritage Command, NH100996)

United Fruit Company's *Atenas* is another example of attempts at reversing perspective. With the amount of smoke belching from her funnel, it is doubtful whether any low-visibility scheme could have provided concealment. (US Naval History and Heritage Command)

Zirkel on 25 September 1918 during trials after completion at the Moore Shipbuilding Co., Oakland, California. (US Naval History and Heritage Command, NH65052)

Also on her trials is *Agawam*, built by the Submarine Boat Corporation at the Newark Bay Shipyard, New Jersey. (US National Archives, 165-WW-273-046)

The dazzle design of *Aroostook* at the Boston Navy Yard, 7 June 1918, has created the illusion of a knuckle running the length of her hull. (US Naval History and Heritage Command, NH57692)

Cargo ship *Volunteer* in San Francisco Bay. (US Naval History and Heritage Command, NH780)

Emergency Fleet Corporation's *Liberator*, 2 July 1918. (US Naval History and Heritage Command, NH102006)

The tones painted on the bow area of *Lake Dymer*, photographed on 17 September 1918, hint of the colouring recommended by Thayer for low visibility. It is in a style that was later evident on vessels painted overall in the US Navy's Measure 16 during the Second World War. (US Naval History and Heritage Command, NH101996)

A distinctive style of disruptive dazzle adopted for a number of ships was this splintered design of triangular areas in contrasting tones, as shown on another Emergency Fleet Corporation cargo vessel, *Andra*, seen here on her trials, 18 August 1918. (US National Archives, 165-WW-505A-003)

Another extraordinary dazzle design, which was the brainchild of American camoufleur Frederick Judd Waugh. It is so effective in its visual distortion that the bow of USS *West Mahomet* appears to be twisted to port. (US National Archives, 19-N-1733)

Western Comet in ballast showing a further example of the Type 9, Design K block or solid geometric dazzle. (Roy Behrens)

Western Comet down in the water, demonstrating that the block geometric scheme was far less effective when a vessel was heavily laden.
(Roy Behrens)

Another example of a block geometry dazzle design, slightly different to the Type 9, Design K, as painted on *Wakulla* in June 1918.
(US Naval History and Heritage Command, NH65036)

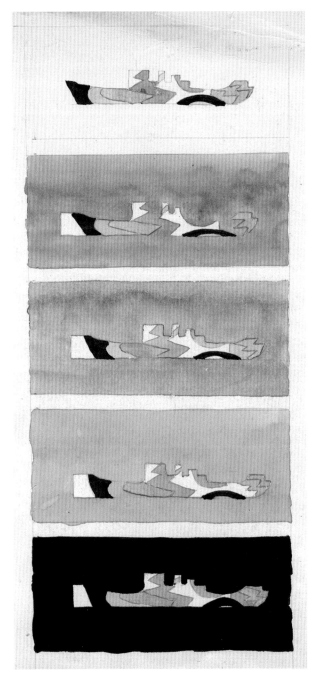

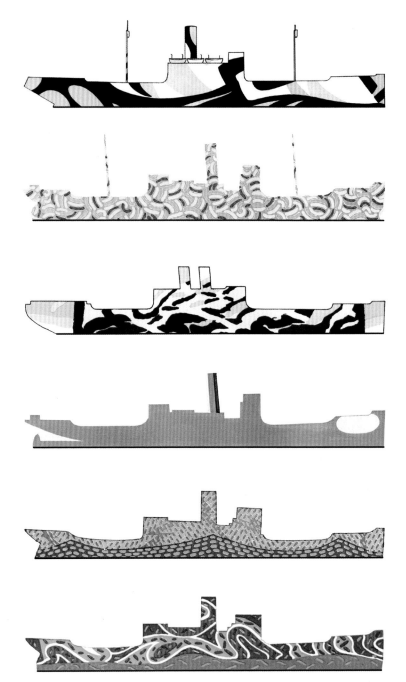

One of numerous painted studies executed by Abbott Handerson Thayer to demonstrate how ship colours can be 'lost' against a complementary background, causing structural disintegration. Having advocated 'ruptive' background blending as part of his theories, it is interesting to note that in this illustration, the profile images have a fairly obtrusive disruptive pattern. (A.H. Thayer estate)

On the request of the Submarine Defense Association, Loyd Jones at the Eastman Kodak Physics Department tested the designs shown here for their visibility coefficient. From the top: Warner Dazzle, Herzog Low-Visibility Dazzle, Toch Low-Visibility Dazzle, Brush Counter-Shading Low Visibility, Mackay Low Visibility and Mackay Low-Visibility Dazzle. The Toch scheme, as shown here, makes an excellent example of hull foreshortening. (Submarine Defense Association, courtesy of Roy Behrens)

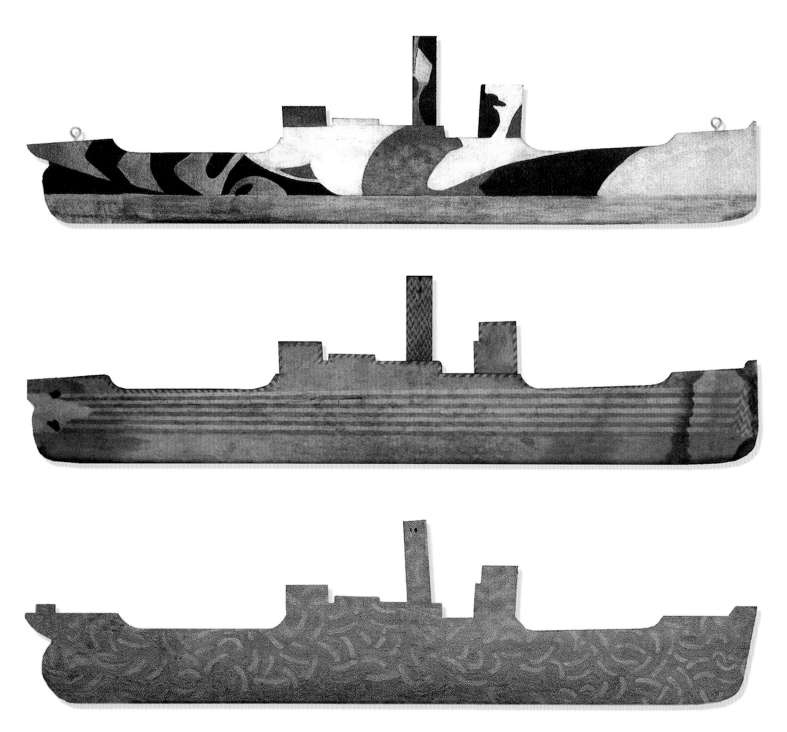

Loyd Jones and Eastman Kodak physicists conducted other tests of colour behaviour in different conditions of natural light on the shores of Lake Ontario, near Rochester. These are examples of some of the painted wooden ship profiles that were studied in these experiments. (Eastman Kodak, courtesy of Roy Behrens)

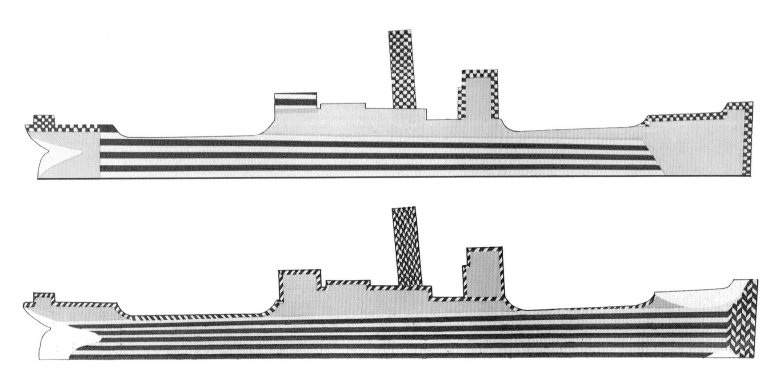

The Submarine Defense Association low-visibility deception schemes recommended (top) for southern latitudes, for use by vessels operating in the Atlantic south of latitude 45 degrees north, and (bottom) for northern latitudes, for use by vessels operating in the Atlantic Ocean north of latitude 45 degrees north. Their subdued disruptive patterns, intended to cause deception at close range, comprised stripes or bands and chequered shapes, which were to be adapted according to the individual lines of a ship, accounting especially for the curvature of the bow and stern. (Submarine Defense Association, courtesy of Roy Behrens)

Six of the dazzle designs created by the Royal Academy team at Burlington House. Note that they are annotated 'Ministry of Shipping Transport Department'. (All courtesy of US National Archives)

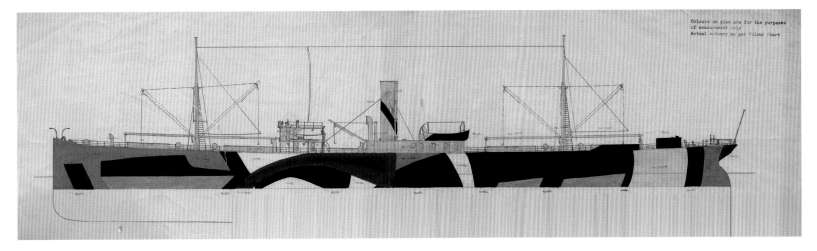

Type 1, Design B, port side.

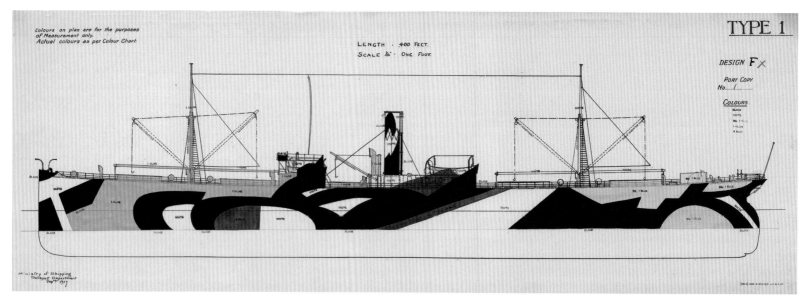

Type 1, Design FX, port side.

Type 2, Design A, port side.

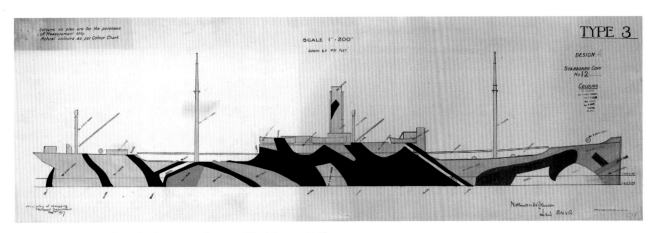

Type 3, Design A, starboard side, personally signed by Norman Wilkinson.

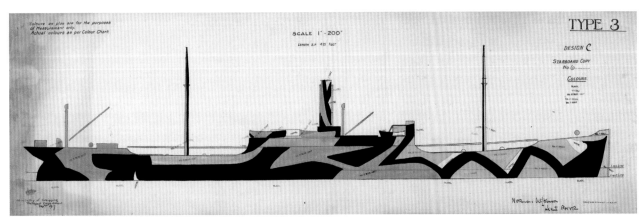

Type 3, Design C, starboard side, personally signed by Norman Wilkinson.

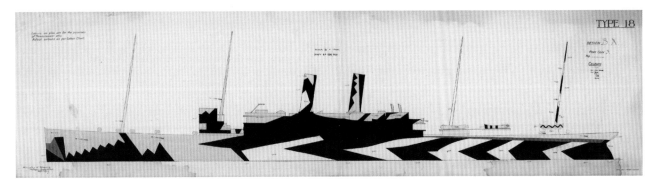

Type 18, Design BX, a design with a pronounced perspective-altering feature.

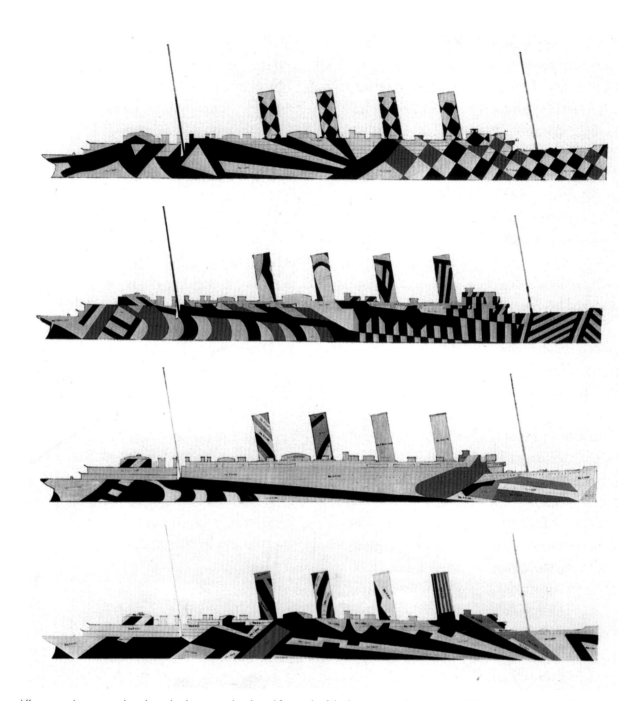

Individual designs, different on the port and starboard sides, were developed for each of the large ocean liners adapted for troop carrying – *Olympic*, *Aquitania* and *Mauretania*. The intention was that their dazzle coats should not only make them difficult to attack successfully but also conceal their identity. This was hardly possible for *Mauretania* (top), with her unique and distinctive chequerboard arrangement, whereas *Olympic*'s dazzle scheme went through several iterations to avoid her being recognised. As explained by Jan Gordon, *Olympic*'s design (second from top) comprised a false bow plus elements to cause confusion and distortion of the bridge and the long lines of the hull, and to counteract the natural curvature of the hull. The bottom image shows that reversed perspective was a stratagem also pursued by the Dazzle Section at Burlington House. (Eileen Tweedy, Crown Copyright)

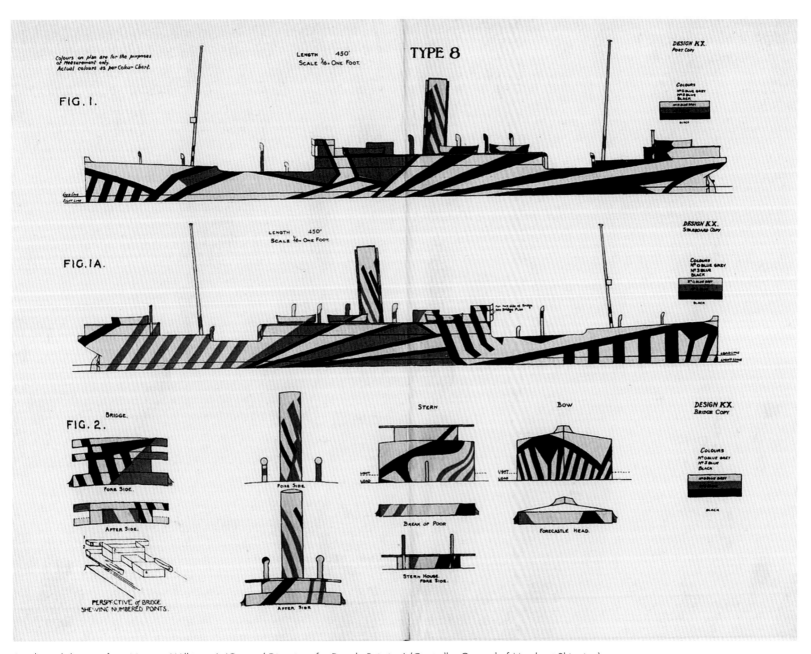

A coloured drawing from Norman Wilkinson's 'General Directions for Dazzle Painting'. (Controller General of Merchant Shipping)

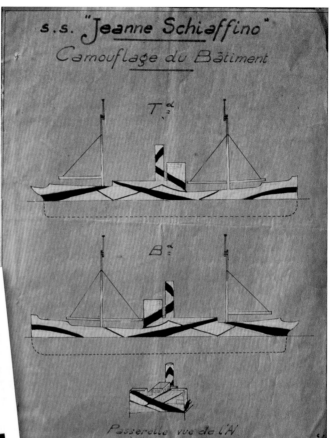

s.s. "Jeanne Schiaffino"
Camouflage du Bâtiment

Passerelle vue de l'AV

Appreciation of the colours of First World War schemes largely depends on contemporary paintings and posters. In this case, the illustration by Arthur Triedler for 'Shoot Ships to Germany' shows a typical Warner disruptive design. (Roy Behrens/Library of Congress)

Colouring plans for SS *Jeanne Schiaffino*, conceived by the French Naval camouflage group at the Jeu de Paume studios. They bear no date but confirm that, like the British and US, the French also reduced their palette of colours to blue and grey tones plus black and white. (Jean-Yves Brouard collection)

EVERYDAY ENGINEERING MAGAZINE

March, 1919
Vol. 6 No. 6

Fifteen Cents a Copy

SCIENCE of CAMOUFLAGE EXPLAINED

Another painted illustration, this one by Howard Vachel Brown, a First World War camoufleur himself. Commissioned for the front cover of the March 1919 issue of *Everyday Engineering*, it shows the Shipping Board model-viewing theatre of the New York district. (Roy Behrens)

Besides the Royal Academy design drawings, the US National Archives and Records Administration also holds an extensive collection of US Navy Dazzle design plans, some examples of which follow. (All US National Archives, courtesy of Roy Behrens)

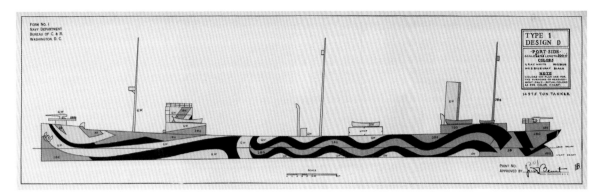

Type 1, Design D, port side.

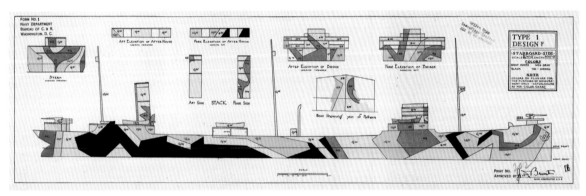

Type 1, Design F, starboard side.

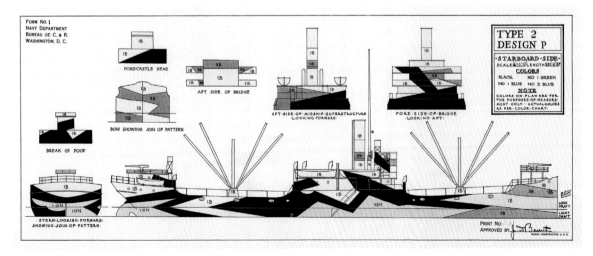

Type 2, Design P, starboard side.

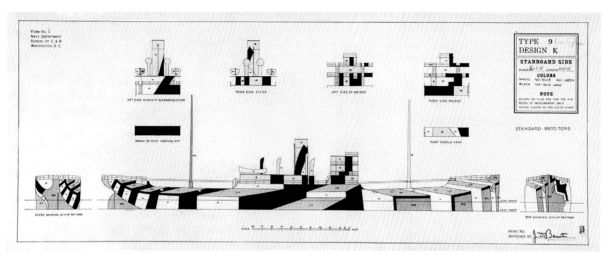

This important dazzle plan from among the US National Archives collection shows the US Navy Type 9, Design K, solid geometry design developed by Everett Warner's Camouflage Sub-Section, which was applied to USS *Isanti* and other ships. (US National Archives, courtesy of Roy Behrens)

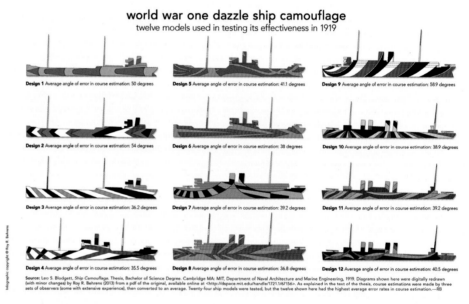

A composite illustration showing the twelve different dazzle designs which were scrutinised by three independent observers for Leo Blodgett's thesis study. (MIT)

An example of a 'Report on Camouflaged Ships', compiled by observers from the US Navy's Office of Naval Intelligence. The artist Thomas Hart Benson, attached to the office, produced this sighting report of the British cargo ship SS *Alban* on 30 October 1918. (Roy Behrens)

Preserved at Halifax, Nova Scotia, this is the Royal Canadian Navy sloop HMCS *Sackville*, repainted in the wartime two-colour (blue and white) Western Approaches camouflage devised by Sir Peter Scott. (Bjarne Pettersen)

Another Canadian warship, HMCS *Restigouche* (formerly HMS *Comet*), is painted in the three-colour (WAG green, WAB blue and white) Western Approaches scheme. (Library and Archives, Canada)

HMS *Belfast*, now preserved on the River Thames, has been repainted in a two-colour Admiralty Disruptive design, not in the Western Approaches scheme as has been suggested. (Dr Allan Ryszka-Onions)

The Polish destroyer ORP *Blyskawica*, preserved in Gdynia, Poland, has been repainted in various of the wartime colour schemes she wore for different operational zones. This first view shows her in June 2008, in what is thought to be a variant of the Western Approaches scheme, with a small area of green paint aft. (Marek Twardowski)

ORP *Blyskawica*, this time in February 2012, in the Admiralty Light Disruptive Pattern for A-class as worn on the Iceland patrol in 1942.
(Marek Twardowski)

German mine barrage breakers moored at Kiel, painted in disruptive camouflage schemes. On the right, in four-colour dazzle, is *Ingrid Horn* (*Sperrbrecher 25*); on the left is the *Minerva* (*Sperrbrecher 13*) in a black and white dogtooth pattern. (Edward Wilson collection)

The battleship USS *Alabama* at anchor in Casco Bay, Maine, in 1942, in US Navy Measure 12 modified camouflage. (US National Archives, 80-G-K-443)

Painted in the disruptive scheme of Measure 32, Design 1D, which she wore from June 1944, is the battlecruiser USS *Alaska*. (US National Archives, 80-G-K-5580)

Another example of the US Navy's patterned disruptive schemes, Measure 32, Design 22D can be seen on the Atlanta-class light cruiser USS *Flint* in March 1945. Beyond her are camouflaged aircraft carriers of the Essex-class. (US National Archives, 80-G-K-3813)

Escort carrier USS *Santee*, seen in one of very few colour views showing the Measure 17 Dazzle System. Her generally muted colours would indicate this to be an unobtrusive variant of disruptive painting. Later, she was repainted in disruptive Measure 32, Design 11A. (US National Archives, 80-G-K-437)

The Aircraft carrier USS *Hornet*, photographed after December 1943 when she was repainted in Measure 33, Design 3A. This and the other wartime colour photographs of naval ships reveals how the colours employed were neutral in tone but relatively high in contrast. (US National Archives, 80-G-K-14466)

The amphibious assault vessel *LCS (L) 11* at Okinawa in April 1945 demonstrates the value of this Measure 31 background-blending design, the colours being mainly greens and browns. (Jean-Yves Brouard collection)

Steaming off the island of Iejima, Motobu, Okinawa, the destroyer USS *Isherwood* shows off her Measure 31, Design 16D camouflage in April 1945. (US National Archives, 80-G-K-4732)

Another amphibious vessel constructed for the island-hopping campaign of the Pacific War is *LSM-152* in a quite different Measure 31 design. Note how on the forward hull the paint has been applied in pointillistic style, like the mottled low-visibility scheme conceived by William Andrew Mackay. (US National Archives, 80-G-K-14463)

USS *West Point*, the converted US Lines passenger liner *America*, seen arriving at New York on 11 July 1945 with repatriated American troops from Europe. Although the war had ended, she still wears her Measure 32 disruptive camouflage. This colour photograph also shows clearly how the colours of this scheme and certain other US Navy disruptive measures are subdued and neutral, whereas in mono photographs they tend to appear bright with greater contrast. (US National Archives, 80-G-K-5776)

The preserved US Navy battleship *North Carolina*, at Wilmington, North Carolina, repainted in her wartime colours of Measure 32, Design 18D. Blue-greys are the dominant colours of the scheme. (Naval Historical Association)

Another preserved US warship from the Second World War. The destroyer escort USS *Slater* at Albany, New York, has been repainted in her Measure 32 modified Design 3D wartime colours, also predominantly of shades and tones of blue. (Destroyer Escort Historical Museum)

5

HYBRID SCHEMES AND EXPERIMENTAL ANTI-RANGEFINDING MEASURES

It may be construed that the ship camouflage schemes which emerged in the USA with the aim of addressing simultaneously low visibility and course deception represented an attempt at mitigation of the antagonism between the proponents of opposing views, but that was never the intention. In point of fact, they were developed from the outset in parallel with the other stratagems as unique practices to provide specific perceptual effects on either side of the blending distance, the point at which pattern colours will appear to become an overall even tone.

It should be stressed that the blending distance is a variable. It is not some constant, precise distance to a point in the ocean from the observer. From a camouflage perspective, it is governed by the granularity of each painted pattern, but it also depends on whether vision is unaided or enhanced by the use of binoculars. Typically, for the schemes categorised here as hybrid, for want of a better term, and intended to induce low-visibility effects at long distance and disruptive effects at closer ranges, it was based on a nominal distance of 5,000 yards, which happened to be the range of U-boat deck guns.

It would be reasonable to argue that some of the schemes described in the previous chapters were also of the hybrid type but the distinction is twofold. The hybrid schemes, as their titles imply, were deliberately conceived to have a dual phenomenon, whereas for those measures designed primarily for the reduction of visibility, unobtrusive deception was claimed as a supplementary side-effect. Moreover, those schemes' attributes, envisaged as causing a disruptive effect at short distances, were not of the dazzle format. In that respect, the only previously described camouflage types that would, perhaps, have qualified for inclusion here were Warner Dazzle and the Submarine Defense Association schemes. The former incorporated certain low-visibility characteristics while the latter exhibited minimal course-deception traits.

Of the eleven schemes subjected to formal appraisal, three can be classified as hybrid, each identified as low-visibility dazzle, that is, presenting particular design qualities intentionally devised to cause a fusion of objectives. On the face of it, such hybrid characteristics are a contradiction in terms, as true dazzle schemes tended to increase visibility rather than reduce it, as already explained.

The originators of these composite camouflage devices were Louis Herzog, Maximilian Toch and William Andrew Mackay, the latter apparently seeking an alternative to the ideas embodied within his original, patented Low-Visibility scheme. His Low-Visibility Dazzle revealed a combination of design elements at considerable variance to the earlier concept, which had been based on the pointillistic integration of primary colours. It was argued, though, that rather than being novel, Mackay's Low-Visibility Dazzle amounted to no more than a mutation of his original philosophy in recognition of the need to constantly reflect on and refine camouflage ideas. Indeed, it was suggested that all five of the camouflage measures first approved for use by the US Treasury Department in October 1917 were in a constant state of flux, such that their approval had been based on their broad concept rather than on the specifics of a particular individual design. That notwithstanding, when Loyd Jones' team at Rochester tested the schemes for their visibility coefficients, Mackay's Low-Visibility and his Low-Visibility Dazzle were treated as two distinct, stand-alone concepts, each having a uniquely different pattern structure and colour mix.

On comparing the three hybrid camouflage schemes, it can be seen that they revealed fundamental conceptual differences. The scheme proposed by Louis Herzog, an artist from New York who was later attached to the US Navy Camouflage Design Sub-Section, was designed to achieve its combined effects by the application of three carefully selected short wavelength colours, i.e. at the blue/violet end of the spectrum. It was claimed that, in juxtaposition, these colours would present as a low-visibility flat tone when they merged at a distance, 'as

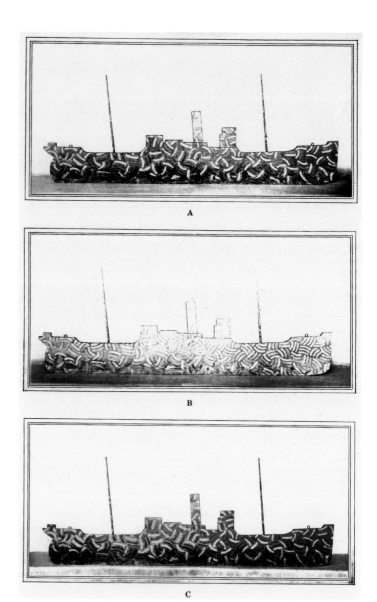

A

B

C

Herzog's Low-Visibility Dazzle scheme was represented pictorially in different tonal interpretations as shown in this illustration of the profiles tested for their visibility coefficient. The colours were mixed tones of blue and green painted in curved segments over a violet-grey base. Like Mackay's Low-Visibility scheme, the pattern distribution had a high level of granularity. Despite the fact that reports state that the Herzog scheme was applied in practice, no examples of ships so painted have been identified. There were either very few, or none at all, the painting complexity perhaps deterring ship owners from selecting this camouflage option. (Eastman Kodak)

a tone scintillating with the atmosphere and blending into it'. Similarly, to cause confusion in determining a ship's heading at closer range, they would tend to shimmer or vibrate, creating an illusion like heat haze. 'Diffusion and interference of light rays, setting up vibrations' was the way Herzog put it. To create these effects, the aim was to break up all horizontal and vertical lines using small concentric arcs of two primary colours painted over a third, base hue. Herzog's view was that 'the eye is more confused by use of circular forms'.

Counter-shading also featured among the ingredients of Herzog's scheme, although it was not explicitly called that. He said, 'All highlights are reduced and the illumination of shadows is given due consideration.' Abbott Handerson Thayer would surely have approved.

Uniquely, in a most progressive consideration, Herzog's proposals were the first to make provision for protection from the risk of aerial attack. To render decks and horizontal surfaces less visible, he proposed that they should be coated in a mosaic of diamond- or lozenge-shaped motifs which would correspond to the appearance of ocean waves when viewed from above. These figures were to be coloured, like the curved panels on the ships' sides, in 'shades and hues of deep violet, blue and green, thus rendering the vessel more nearly the colour of the waters'. As a paradigm of background blending, this too would have received Thayer's full endorsement.

Herzog's Low-Visibility Dazzle had achieved a relatively low visibility rating in the Eastman Kodak tests, but its dazzle credentials were less convincing and it is questionable whether they would have evoked the desired course deception effect. More likely, it was not dazzle in the sense of disruption of perspective but intended instead to be dazzling to the eye, as in to 'glisten' or 'glare'.

As for Maximilian Toch's concept, it is evident that his scheme did evolve. Descriptions on contemporary photographs refer to 'Original Toch' and 'Improved Toch'. As an industrial chemist and paint manufacturer, also residing in New York, Toch had previously formulated the paint mixture for the US Navy's Standard Grey. After the formation of the US Navy Camouflage Section, he was attached to the organisation's subordinate Research Sub-Section at Rochester.

With his scientific background, it is likely he brought a systematic approach to camouflage development and, perhaps as a consequence, his proposed hybrid system was quite different to that of Herzog. It comprised large, abstract areas or diagonal streaks of contrasting colours, whose chromatic properties would be conducive to blending into felicitous tones at a distance, with darker shades concentrated in the lower hull near the waterline and lighter shades to a greater extent in the upperworks. While the pattern complexity and detail of the Herzog scheme was close in appearance to Mackay's original scheme, the Toch

One of two interpretations of the original Toch Low-Visibility Dazzle scheme (see the colour section for the alternative design). It comprised large, abstract areas or diagonal streaks of contrasting colours – black, white, grey, dark green and light pink-purple – whose chromatic properties were said to be 'conducive to blending into felicitous tones' at a distance, with the darker shades concentrated in the lower hull near the waterline and the lighter shades, to a greater extent, in the upperworks. Dark green was avoided in the superstructure, while for funnels and masts, a combination of light blue-grey with splotches of dark grey and light purplish-pink was recommended. (Eastman Kodak)

designs bore a stronger resemblance to those of Everett Warner's Dazzle and, in some respects, they hinted visually of John Graham Kerr's Parti-Colouring concept. Contemporary photographs reveal that the pronounced bias towards hull foreshortening devices shown in Toch's working sketches (see colour illustrations) was noticeably absent in the examples of his system as it was actually utilised.

Essentially, William Andrew Mackay's Low-Visibility Dazzle was also at variance from its hybrid competitors. Equally, it was not at all like his Low-Visibility system, reinforcing the view that it had not evolved from the earlier proposal but stood alone as a novel camouflage option in its own right. Vivid colours – green, orange and white – more like those first advocated by Wilkinson and Warner, characterised this camouflage variant. They were painted over the entire hull and upperworks in bold,

undulating, even snake-like shapes above a bright blue band along the length of the hull above the waterline. In the absence of any recorded pronouncement, it can only be assumed that official endorsement was granted to William Andrew Mackay's revised scheme; even so, just a single example of it, significantly modified, has been identified.

While Mackay's working methods are to some extent known, this is not so for Herzog and Toch. However, the fact that little is known of how they arrived at their camouflage concepts is of little consequence, since all these hybrid schemes were prematurely killed off, along with the other pre-dazzle devices, by the US Navy Department's unilateral declaration of January 1918.

★★★

It is believed that the Mackay Low-Visibility scheme was slightly modified as it appeared on USS *Isabel*, with the pattern components contained within a grid structure. It may be that this represented a step in the evolution of Mackay's thinking towards Low-Visibility Dazzle. (US Naval History and Heritage Command, NH544)

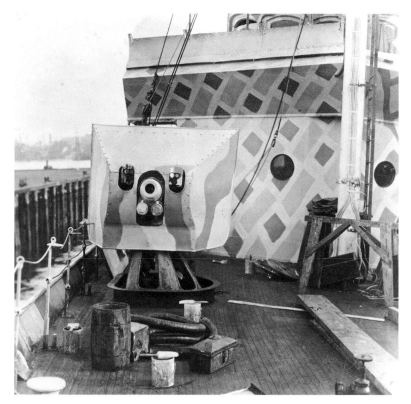

In this close-up view of USS *Isabel*, the grid-line detail can be seen more clearly. Both photographs of *Isabel*, a private yacht taken over for escort duties, were taken at the Boston Navy Yard, Charlestown, Massachusetts, on 31 December 1917. (US Naval History and Heritage Command, NH545)

The only identified example of Mackay's Low-Visibility Dazzle is said to be as seen here on *De Kalb* at the Philadelphia Navy Yard on the Delaware River on 18 February 1918. In fact, the scheme depicted could be a fusion of the Low-Visibility and Low-Visibility Dazzle variations because the solid blue panel along the lower hull of the latter design is missing. Instead, low down there is the pixelated pattern arrangement common to Mackay's Low-Visibility. (US Naval History and Heritage Command, NH54662)

A derivative of Mackay's Low-Visibility scheme on the converted steam yacht USS *Venetia* at Mare Island Navy Yard, California, in October 1917. (US Naval History and Heritage Command, NH47198)

Battleship USS *New Jersey* during the US Navy's Fleet System anti-rangefinding trials, also showing a variation of Mackay's Low-Visibility scheme. (US Naval History and Heritage Command, NH100409)

Painted in the Toch scheme. USS *Aztec* at the coaling facility, Boston Navy Yard, on 20 December 1917. (US Naval History and Heritage Command, NH543)

USS *Wilbert A. Edwards* at the Norfolk Navy Yard on 18 August 1917, showing a variation of the Toch scheme, painted according to a Navy General Order issued in July 1917. (US Naval History and Heritage Command, NH75514)

The Toch scheme had several variants. Here USS *Amagansett* is seen in Toch Original at the Norfolk Navy Yard, Portsmouth, Virginia, in September 1917. An appended note suggested some work carried out, possibly the paint job, had cost $461. (US Naval History and Heritage Command, NH100664)

USS *Warren J. Courtney* was painted in a Toch scheme variant after the Navy General Order of July 1917. (US Naval History and Heritage Command, NH75516)

USS *Kajeruna* painted in what was described as the Toch Improved scheme, a more rudimentary design arrangement. (US Naval History and Heritage Command, NH101951)

The US camouflage measures so far discussed in this and previous chapters all primarily pertained to merchant ship protection. But the issue of painted defensive measures for warships of the battle fleet remained open. In the UK, Kerr had unsuccessfully proposed parti-colouring as an anti-rangefinding practice, while plain grey alone had been deemed inadequate and dazzle was declared inappropriate for large naval vessels like battleships and battlecruisers, since their course bearing could be readily deduced from observation of their movement as an integral squadron.

Yet there were those who remained of the opinion that some sort of painted protection was called for to render capital ships less vulnerable to surface bombardment by interfering with the operation of fire control rangefinders. As a result of pressure to develop such measures, a series of what were called the 'Fleet System' trials were initiated by the US Navy through the winter of 1917 to 1918, exploring various deception artifices.

In fact, the British had made one final attempt at something of the kind when, in 1918, it tried a type of anti-rangefinding black and white dazzle painting on the battleships HMS *Revenge*, *Royal Sovereign* and *Ramillies*, with a more subdued version on the battlecruiser HMS *Repulse*. However, the tests concluded that inhibition of target range determination by any type of rangefinder, whether coincidence or stereoscopic, was unachievable, ending all British initiatives in that regard. The only recommendation was for top masts, which could assist rangefinding, to be reduced in height because wireless improvements had made reliance on such lofty aerials redundant.

The US authorities went to somewhat greater lengths in their quest for a type of anti-rangefinding camouflage, although, in the final analysis, they arrived at the same outcome as the Admiralty. Rangefinding, as the means of setting the angle of fire of deck-mounted guns, relies on the alignment of the horizontal or vertical lines of the target vessel. The distance is then computed by trigonometrical calculations from the instrument readings in much the same way as the system employed in submarine periscopes. The rangefinders on naval surface ships, unrestricted by space considerations, were, however, somewhat larger and more sophisticated.

Numerous interference devices were assessed under the 'Fleet System' trials, among them Mackay's Low-Visibility scheme, various forms of disruptive painting, hull foreshortening and the painting of dogtooth panels, alluding to indentations, along the edges of the weather decks. Dynamic measures also featured, as had also been tried in the UK, in the form of triangular, metal pennants suspended between funnels and masts.

Possibly the most extraordinary anti-rangefinding scheme proposed, which may or may not have been part of the 'Fleet System' exercise, emanated from the Norfolk Navy Yard at Portsmouth, Virginia. Devised by Frank Morris Watson Sr, a master painter at the shipyard, it was characterised by bold, symmetrical, diagonal or rainbow-like geometric patterns painted the length of the ship, over hull, turrets and upperworks, in strident, highly contrasting tones. Fanciful interpretations of these schemes have suggested that the colours used were purple, dark green and white, but it is now thought that in all probability they comprised a combination of black, white and two or three intermediate grey shades. Obtrusively disruptive in their properties, these schemes stimulated profound perspective distortion and, although intended for warships, may have performed just as well as course confusion measures for more general application.

After expending significant time, money and effort on these experiments, the 'Fleet System' and Norfolk Navy Yard trials were wound up and nothing more came of them. The reality was that no definite case existed by that time for anti-rangefinding protection for capital ships. US battleships never engaged in any surface action during the First World War and, after Jutland, no further encounters of any significance took place between the British and German battle fleets.

As for Frank Watson Sr, who hailed from Philadelphia, he continued as leading painter at the Norfolk Navy Yard and, in 1941, over twenty years later, it remained his occupation.

To break up horizontal and vertical lines, another experiment involved painting an array of saw-teeth shapes along the top edge of the hull at weather-deck level and placing triangular metal pennants on or between the lattice masts, as seen here on USS *Utah*. (US National Archives, 19-N-2034)

USS *Rhode Island*, at the Boston Navy Yard, was experimentally painted in a form of disruptive dazzle. (US Naval History and Heritage Command, NH101192)

USS *Montana* was also treated to an experimental variation of disruptive dazzle. (US Naval History and Heritage Command, NH73605)

USS *Florida* in an experimental camouflage scheme at the Norfolk Navy Yard, Virginia, on 23 November 1917. Confined mainly to her funnels, the painted-out section at the bows is very effective in causing foreshortening. (US Naval History and Heritage Command, NH78794)

USS *Nevada* had illusory indentations painted down her stem in addition to the saw-teeth shapes. During this phase of the evaluation exercise, a third battleship, USS *Oklahoma*, also had disruptive dazzle added to her funnels. (US National Archives, 165-WW-334-040)

USS *Nebraska* painted in Watson-Norfolk Design No. 3. Photographs suggest she was painted the same on both sides. The pattern exhibits strong false-perspective effects, especially at the stern which appears to be curved away. (US Naval History and Heritage Command, NH101208)

USS *Montgomery* in Watson-Norfolk No. 3 on her port side. She is in dock at the Norfolk Navy Yard on 1 February 1918. (US Naval History and Heritage Command, NH75517)

Another view of *Montgomery*'s starboard-side Watson-Norfolk No. 4 camouflage design, taken when she was off New London, Connecticut. (US Naval History and Heritage Command, NH57227)

Opposite: Probably taken at the same time, this view shows the starboard side of *Montgomery* painted in Watson-Norfolk No. 4. It is not known how many ships were experimentally adorned in the Watson-Norfolk scheme, but presumably there were also vessels painted in Design Nos 1 and 2. (US National Archives, 165-WW-70G-026)

Visual Tricks 2

THE 'HOPEFUL HOAXES'

Another group of visual tricks, dubbed the 'hopeful hoaxes' because their ability to deceive was deemed dubious, sought not to imply the absence of some part of a target vessel but, in effect, the complete opposite – the presence of something which did not exist. The effectiveness of these devices was questioned by seasoned and experienced seafarers who had doubts as to whether U-boat commanders really could be convinced into accepting as genuine the imagery confronting them.

Essentially, the tricks fell into two types, namely fake ships painted on the hull sides and simulated bow waves and other indicators of water disturbance arranged along the waterline. In the first case, the idea was to give the impression that another vessel lay between the target ship and the submarine, partially obscuring it and potentially obstructing the path of a launched torpedo.

In the First World War, the preferred silhouette represented a destroyer, bestowing a dimension of deterrence to the artifice. Destroyers were the U-boats' nemesis, particularly from 1916 onwards when they carried a deadly payload of depth charges and deck-mounted projectile launchers. Nonetheless, the false ship concept had limited application in the First World War and was soon abandoned, although according to H.M. Le Fleming, a simplified derivative was favoured for Royal Navy ships between 1915 and 1917.

The second fake image device, implemented more extensively during the First World War, was the false bow wave, sometimes applied with spurious ripples along the ship's side and even fictitious indication of wash or wake at the stern, all intended to convey the impression of movement when there was none or, if underway, of greater speed than the actual. Having a discernible, though limited credibility, the use of this deceptive practice continued throughout the First World War.

Despite any lack of certainty as to the effectiveness of these particular visual tricks, both were revived during the Second World War, the false bow wave to a greater extent. Fake ships and fallacious bow waves varied considerably in the style of their execution, from the crude and simplistic to the complex and extraordinarily detailed.

The First World War M-class destroyer HMS *Myngs* with false bow wave.
(Adrian Vicary)

The C-class cruiser HMS *Cairo* has an elaborate hoax combination of false bow
wave, false side wash and false stern wake. (Crown Copyright)

HMS *Gloxinia* sports a most flamboyant false bow wave in a view from the Second World War. (Adrian Vicary)

The French called the sweeping white curves of the sham bow wave 'Le Mustache' because of its resemblance to the stylish Edwardian fashion of upper-lip decoration. Here, the suitably adorned battleship *Paris* of the Courbet-class is seen at Plymouth in 1941. She was seized there on 3 July 1940 at the time of the French capitulation. Note the Union Jack flying from her forward flag staff. (Crown Copyright)

The heavy cruiser USS *Northampton* wears a Measure 5 false bow wave in Brisbane, Australia, on 5 August 1941, four months before America entered the Second World War. (US Naval History and Heritage Command, NH95333)

In Brisbane around the same time, USS *Salt Lake City* in an unusual variant of the US Navy's Measure 5, having a dark false bow wave. The reasoning behind this variation is not known, nor do we know whether it was effective for its intended purpose. (US Naval History and Heritage Command, NH80837)

The Japanese seaplane tender HIJNS *Akitsushima* has an extraordinarily diverse disruptive camouflage scheme. Besides the evident false bow wave and stern wake, she has striped paint forward and patches of dappled paintwork on her bow, side, funnel and derrick. (Shizuo Fukui)

In Norfolk Harbor, Portsmouth, Virginia, on 29 November 1917, USS *Antigone* (formerly Norddeutscher Lloyd *Neckar*) has a ghostly image of a destroyer painted on her hull side. The ruse may have been more convincing in the conditions of murk and mist that are present in this picture. (US Naval History and Heritage Command, NH57625)

Like *Antigone*, another former German liner seized by the USA, USS *Von Steuben*, is seen here with a fake destroyer on her starboard side. The silhouette, complete with its own bow wave, was also painted on her port side. (US Naval History and Heritage Command, NH101626)

The former Italian cruiser *Marco Polo*, renamed *Cortelazzo*, has a more ambitious fake decoration comprising two false ship images, presumably to better synthesise realism. (Ufficio Storico)

The small passenger ship *Chella*, of La Compagnie de Navigation Paquet, at Marseilles while under attack on 2 June 1940. The false craft painted on her side is thought to be unique, being three-dimensional, suggesting a vessel that is underway and headed in the direction of the observer. The fake image is unusually complex and even has realistic hull colours. (Jean-Yves Brouard collection)

This unidentified German *sperrbrecher* (mine barrage breaker) has two false ships painted on her port side, both bristling with guns. It is believed that this camouflage ruse was short-lived on German naval auxiliaries since the majority were repainted in extreme disruptive schemes of the '*flimmestarnung*' type (see the lower illustration on page 164). (Edward Wilson collection)

A quite different approach to the false ship illusion is seen on the French battleship *Richelieu*. Instead of an image depicting an entire ship, she has just the bow of another ship on her side with its fake bow wave. Later, she was repainted with a number of these bow shapes along her length to give the impression that she was underway with several escort vessels close alongside her. In the Admiralty Standard (Light Tone) schemes of 1944 onwards, the simplification went even further, amounting to no more than a dark panel painted amidships. (US National Archives, 80-G-78789)

6

EVALUATION – ENQUIRIES, STUDIES, OPINIONS AND STATISTICS

The subject of First World War marine or ship camouflage continues to elicit interest, discussion and debate, even more than 100 years after the event – both the question of its effectiveness and the specific details of individual schemes or designs. It is often claimed that rigorous, analytical assessment as to the effectiveness of these early painted protection measures was generally lacking and studies were inconclusive.

Although it is correct to say that much of the appraisal was anecdotal and subjective in character and potentially influenced by the prevalent conflict of opinions, some rigorous, quantitative investigation and analysis did take place, which reached important and definite conclusions. Nonetheless, during the inter-war period, the unsubstantiated conclusions of earlier official inquiries were often repeated and there was a tendency, without justification, to side with either the low-visibility or disruptive-painting positions or the idea of one camouflage proponent over another.

The opportunity arose, as the First World War drew to a close, to properly appraise the various camouflage schemes that had been promulgated over the four years of war. While the primary purpose of the exercise was to determine what had or hadn't worked, in truth, it was only Dazzle and disruption that were singled out for critical evaluation. It is a reasonable question to ask, given the conflict, even animosity, between the proponents: why was it that the low-visibility concepts were not subjected to the same level of scrutiny? Indirectly, the process of evaluation triggered, in the UK, an acrimonious dispute between two of the main proponents when an inquiry considered who was to receive recognition as the inventor of Dazzle. The aim here is to review the available evidence pertaining to dazzle disruption by looking at the results of the various inquiries and studies, regardless of their conclusions, taking into account collateral factors that could have had a bearing on the outcomes of the deployment of these practices.

Much of the ship loss data collected at the time is patchy, inconsistent and lacks key information concerning the circumstances of the loss by which it could be analysed fully, impartially and objectively. In part, these shortcomings were unavoidable because of the war conditions. Communication difficulties and security restrictions would have made it difficult to gather sufficient data in a controlled manner in accordance with defined parameters. Even so, when conclusions were drawn, little consideration was given to extraneous factors that may have influenced the scanty statistics gathered – the sizes and types of the ships concerned, whether or not they were sailing alone or in convoy, the visibility conditions, the time of day, the sea area and which system of camouflage they were painted in.

The British authorities were, it is suggested, circumspect about the efficacy of dazzle although their deductions were, if anything, ambivalent. In contrast, despite opposition from some quarters, in the USA and France it was enthusiastically endorsed, which raises another question – why should there have been such disparate verdicts?

It is not the purpose of this book to take sides in the concealment versus disruption debate, only to present such factual evidence as can be found in official statements and study reports. However, in order for this volume to provide a balanced account of the story of marine camouflage, instances of illogicality should be mentioned. It is evident, for instance, that Admiralty judgements vacillated at different times, with no apparent rationale, often contradicting themselves. Clearly, there was an overriding need to have the evidence interrogated independently, dispassionately and free from service attitudes. Fortunately, as will be seen, one comprehensive evaluation of dazzle, unfettered in any way by officialdom, did take place.

The small amount of numerical data for ship losses (sinkings and extreme damage) that was compiled by the Allied naval authorities is summarised as follows:

	Total Ships Dazzled	Total Ship Losses	Dazzled Ship Losses
Admiralty (UK)	2,367 (May/June 1917–30 June 1918)	1,075 (1 Jan– 30 June 1918)	478 (1 Jan– 30 June 1918)
USA	1,127 (1 March– 11 Nov 1918)	96 (1 March– 11 Nov 1918)	18 (1 March– 11 Nov 1918)
France	1,500 (July 1917– Nov 1918)	– (included above?)	– (included above?)
TOTALS	4,994	1,171	496

Notes:

1. It has been reported that a total of 4,000 merchantmen and 400 warships were dazzled in British schemes alone, but these numbers cannot be substantiated.

2. The figures for total ship losses and dazzled ship losses for the Admiralty are, as far as is known, a combination of British and Allied casualties, including French flag but excluding US flag losses.

3. The 18 US flag losses include 4 that were collision casualties and three that were mined, leaving just eleven caused by enemy torpedo attacks.

These figures, inadequate as they are, reveal that only 11.5 per cent of US total losses were dazzled ships. They also show that less than 50 per cent of all Allied ship losses were vessels on which dazzle had been applied, but this is only part of the story. Figures for attacked ships which were damaged but not total losses marginally favoured dazzled ships in the ratio of 3 to 2.4. But, without trawling through thousands of logs of surface ships and U-boats, there is no way of knowing how many dazzled ships were subjected to torpedo attack where the weapons missed the target, which would have supported the claim that dazzle hampered successful attacks.

The huge difference in the reduction of casualties, as a possible benefit of dazzle, between US-registered ships on the one hand and British and Allied ships on the other is extraordinary. As has been stated, there are numerous factors that have a bearing on this data, one of which could be the difference in number of US ships that were operational at the time compared with those operated by the other Allied nations. Another factor could have been greater effectiveness of the US dazzle designs as time went on.

Official inquiries were instigated in the UK, USA and France and it is appropriate to consider the methodology applied (if any) in each case, how reliable it was, from where the data was collected and what conclusions were derived from it. The Admiralty inquiry into the value of Dazzle painting, commissioned in the spring of 1917, was to some extent motivated by First Lord of the Admiralty Sir Eric Geddes, who was an outspoken critic of dazzle disruption, holding the view that it was a waste of money that would be better spent on other naval equipment needs. The inquiry examined shipping losses of dazzled and non-dazzled ships and reviewed anecdotal observation reports from ships' masters, both naval and commercial.

Although some reports provided information on the weather and viewing distance relevant to observations, in the absence of a formulated requirement for this type of information, it was often omitted. The data for losses and damage cases was recorded, for merchantmen, by Lloyds of London and, for naval vessels, by the Admiralty, cross-correlated to the Dazzle Section's records of dazzle-painted ships.

Extracts from the conclusions of the Admiralty inquiry, published in July 1918, have been quoted repeatedly, but usually selectively, giving the impression that support for the continuation of dazzle was reluctant and only for reasons of sustaining crew morale and that, otherwise, the 'wholesale' use of the system was opposed. Most often, the following, overtly critical paragraph of the report is restated:

> It must be remembered that the sole object of dazzle painting is to cause confusion as to the course and speed of a vessel, and that it is not designed to reduce visibility. In our opinion, from a careful examination of the whole of the evidence, no definite case can be made out for any benefit in this respect for this form of camouflage.

However, ambiguously, this statement was preceded by a less-often published paragraph expressing a contrary view:

> It is considered that dazzle painting cannot possibly assist the submarine and it is almost certain to increase the difficulties of attack by making it difficult to tell a ship's course. The report also concedes that 'at the same time, the statistics do not prove that it [dazzle] is disadvantageous'.

The inquiry undertaken by the US Navy, which reported in May 1920, followed along similar lines to that of the Admiralty in that it collated the observation reports of ships' officers, but it also drew on the detailed reports compiled by the observers of the Office of Naval Intelligence using the formulated 'Report on Camouflaged Ships' document, which prompted the provision of additional details. The accumulated statistics for ship losses, dazzled and non-dazzled, wholly supported the US inquiry's deductions: 'It is considered beyond doubt that [dazzle] camouflage painting was of distinct value, particularly in the case of large and fast vessels, which might be saved from disaster by the momentary confusion of the attacking submarine commanders.'

Of the three principal Allied nations that adopted dazzle painting for ship protection, the French attempted to apply certain, more analytical criteria to their method of data collection. As in the UK, there was opposition in the French Navy to the use of dazzle-type painting, notably from Admiral Schwerer, who issued a critical report on 30 April 1918 following a patrol in the Brittany division in which he had been present. Therefore, in order to reach an objective conclusion as to the positive benefits or otherwise of dazzle disruption, the French Naval authorities conducted two parallel studies, as explained by Marc Saibène in *La Marine Marchande Française 1914–1918*.

On 16 May 1918, the government minister responsible for the navy, circulated directives to all maritime authorities, naval and commercial, prescribing the requirements and basis of two official camouflage surveys. The requisite information, including from submarine commanders, was to be channelled via the DGGSM (Direction Générale de la guerre sous-marine).

The first survey was focused on naval escorts and patrol vessels, the second on merchant ships including converted troopships. The French also had maintained meticulous records of the ships that had been dazzled, by type, design and date, all of which were beneficial to the surveys' credibility.

Although there remained an element of conjecture in the feedback received, the supplementary particulars supplied enabled the subsequent assessment to consider, besides any opinions, the related circumstances of most visual encounters. The majority of the responses received emanated from naval officers rather than mercantile master mariners, which may have had some influence on the conclusions. Nonetheless, published on 3 August 1918, the official summary report revealed overwhelming support for disruptive dazzle, some 80 per cent of the testimony expressing a view in its favour. The submission of one naval commander succinctly reflected the judgements expressed by many of his contemporaries:

> The camouflage does not decrease the distance you can see a ship in that her funnel, her masts, her smoke are detected regardless of the method of camouflage, but it undoubtedly increases the difficulty of determining the course both by day and night. One thing is definitely ascertained, [it] can in many cases mislead the submarine, forcing him to extend his periscope examination and be, in a word, hindered.

In its conclusions, while conceding that the dazzle process was not without certain valid criticisms, notably that white paint was too revealing and should be substituted by grey, blue-grey or light olive green, the report recommended that the application of dazzle painting should definitely be continued.

The French inquiry was explicitly directed at the assessment of dazzle's ability to cause course distortion. That of the Admiralty, while ostensibly concerned with the same issue, also considered the system's level of visibility, which may go some way in explaining the contrast in the two outcomes.

Around this time, the waters were rather muddied when another quite different inquiry was launched in the UK that had little or nothing to do with the effectiveness of Dazzle but, rather, was concerned with determining who should be credited with the provenance of the invention. The opposing claimants were narrowed down to three: Lieutenant Commander Norman Wilkinson, Professor John Graham Kerr and Mr Archibald Phillips, a Liverpool art dealer.

One of the contenders for the credit of inventing dazzle painting, as assessed by the Admiralty Committee on Invention and Research, was Percival Tudor Hart, who was responsible for this unusual zig-zag scheme experimentally applied to a naval harbour launch. (Crown Copyright)

A Royal Commission on Awards to Inventors had been constituted on 19 March 1919, chaired by Judge Baron Thomas Tomlin, its task to commence deliberations on numerous claims in respect to wartime creations and innovations. In fact, where Dazzle was concerned, the Royal Commission was preceded by an inquiry for the same purpose by the Lords of the Admiralty, overseen by a committee under Admiral Sir Arthur Farquhar for the Board of Invention and Research, which first sat on 27 October 1919. A month later, on 27 November 1919, each of the three claimants made a submission to the committee.

The essence of Phillips' claim was that, having been inspired by the sight of war shipping in the River Mersey as he had crossed each day to and from work, he had conceived of a scheme in which a unique hexagonal mosaic design in particular colours would both disguise and have the effect of 'dazzling the eyes of the gunners of enemy submarines'. These ideas had been transmitted as a proposal with sketches to Lord Derby at the War Office on 9 May 1915. Just nineteen days later, he likewise sent the same material to the First Lord of the Admiralty. Apart from acknowledgements of receipt, nothing came of his initiative. Later still, in November 1917, when the gaudily dazzled Cunarder *Mauretania* was witnessed during a Liverpool call, he contacted the Admiralty again, partly concerned that his conception had been hijacked but also recommending supplementary measures for use with Dazzle, based on the 'fake ship on the side' stratagem.

When, in October 1920, Kerr was informed by the Admiralty inquiry that they had decided not to rule in his favour, he was advised, if he was not content with the verdict, to pursue his claim further through the Royal Commission. Represented by legal counsel, for a second time each claimant's case was heard by the commissioners on 16 October 1922. Further submissions relating to dazzle painting were made on behalf of the Ministry of Shipping by John Anderson, head of its civil service staff. Much of the proceedings dealt with the semantics of the terms 'dazzle', 'dazzling' and 'dazzled', along with the difference between defence against submarines versus the inhibition of surface attack.

The commission's decision, in favour of Norman Wilkinson, was declared with royal approval on 21 October 1924. Kerr had been rebuffed again while the committee disparagingly dismissed Phillips' concepts as having 'no value at all'.

The commission's declaration that it recognised all three men's aims had been 'governed by a patriotic motive and spirit' did little to appease the disgruntled Kerr and Phillips. While the matter had been resolved, as was thought, from an official standpoint, the claims and counter-claims

A. M. LANGDON,
ROBERT YOUNG.
A. CHASTON CHAPMAN.

P. TINDAL-ROBERTSON
(Secretary).
Dated the 21st day of October, 1924.

APPENDIX I

Name of claimant	Nature of claim	Amount awarded
228. Mr. A. Phillips	Dazzle Painting for Ships (Head III)	Nil.
229. Professor J. Graham-Kerr	Dazzle Painting for Ships (Head III)	Nil.
230. Lieut. Comdr. Norman Wilkinson	Dazzle Painting for Ships (Head III)	£2,000.
231. Lieut. R. Rimington	Improvements in Stokes Gun Projectiles (Head III).	Claim withdrawn at the bearing.
232. Mr. S. C. E. Stone	(a) Cartridge Clips for 6" Trench Mortars	£250.
	(b) An Elastic Attachment Device for Augmenting Rings for Service Gun Ammunition (Head III).	Nil.
233. Major F. V. Lister	(a) 4" Stokes Bomb and Propellant	£500.
	(b) Stokes Bomb Ignition	Nil

Excerpt from the Royal Commission on Awards to Inventors Report No. 3, dated 21 October 1924. Covering claims lodged over the period from 1 August 1922 to 31 July 1924, the third report records the date when the award was officially made to Norman Wilkinson, number 230 in the list. (Crown Copyright/HMSO)

wrangled on contentiously at a personal level until John Graham Kerr's death on 21 April 1957, eighteen years after he had been knighted.

The most informative and dependable study of dazzle painting, completely independent of government involvement, ensuring its impartiality and neutrality, had been launched back in 1919 in the USA, before even the Admiralty Board's inquiry had commenced its first session. It was the first, and probably the only, properly scientific analysis of dazzle disruption from that period, concentrating on the effect of course distortion.

Leo S. Blodgett, a student at the Department of Naval Architecture and Marine Engineering at the Massachusetts Institute of Technology, had chosen for his BSc thesis the topic of 'Ship Camouflage'. For his investigation, he proceeded to carry out the most thorough research, fundamentally governed by an exclusively quantitative treatment of the data gathered through an assiduously controlled method of observation of test models. The scholastic work that resulted, published on 12 May 1919, though less well known, was crucial in affirming the validity of dazzle's effectiveness.

The institute, and Blodgett for that matter, had been the beneficiary of a quantity of valuable equipment donated at the war's end on the orders of Henry C. Grover of the Emergency Fleet Corporation, the main item being the model-testing theatre previously deployed in the Shipping Board's Boston District, as well as a quantity of models and other items contributed by William Andrew Mackay, including original draught designs from the New York District.

The arrangement of the testing theatre and associated equipment, though much less sophisticated, exhibited features in common with those previously constructed at the Royal Academy, London, at the Jeu de Paume, Paris, and in the Washington DC laboratories of the US Navy Camouflage Section. They were, though, adapted in a number of ways for the purpose of Blodgett's thesis.

The optical system, modified to produce images of the models more accurately as if viewed from 1,100 or 2,200 yards, comprised two lenses with parallel mirrors angled at 45 degrees to the focal plane and mounted in a box-like enclosure to simulate a U-boat periscope. The lower lens had a fixed focal length of 30in, the upper, reducing lens, having adjustable focus. In front of and at the bottom level of the

The reconstructed viewing theatre at the Massachusetts Institute of Technology. The surface of the theatre platform was intentionally given a convex curve (i.e. higher in the centre) to create the illusion that the models were floating on a body of water. A selection of sea and sky scenes painted on broad horizontal and vertical canvas belts, in a variety of combinations which were assumed might be expected to occur at sea, were positioned by moving the belts either up and down or backwards and forwards over rollers using hand cranks. The horizontal belt can be seen above the surface, returning beneath the platform. There were four seascapes – calm blue sea, bright green sea with some wave disturbance, dull grey hazy sea and rough, choppy sea with white wave caps. (MIT/Roy Behrens)

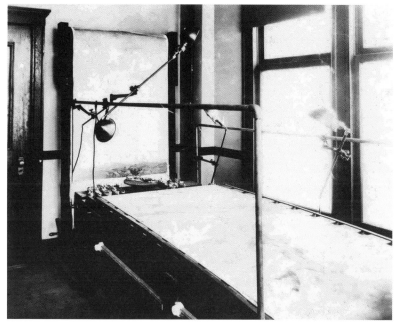

Arranged horizontally along the length of the platform were nitrogen daylight-balanced lamps. The second canvas belt, seen in this view, erected vertically at the theatre's end furthest from the viewer, had a selection of painted sky backdrops on it representing various weather conditions. This, too, could be moved round on rollers to change the skyscape as required for different tests. There were four skyscape options – clear blue sky, hazy sky, dull sky in which cumulus clouds predominated, and dark, cloudy stormy sky. (MIT/Roy Behrens)

periscope was a fog-simulating device comprising a semi-transparent mirror angled vertically at 45 degrees to the line of vision with, above it, a ground glass screen placed parallel to the line of vision. Various conditions of haze, from fog to light mist, could be contrived using a dimmer-controlled light source.

A turntable was installed towards the sky-backdrop end of the platform. This could be rotated using a hand wheel connected to a pointer on a calibrated compass card to indicate the true bearing of each model ship. During observations, to avoid any undue influence, the actual bearing settings were concealed from the viewers' sight.

The bearing angle perceived by the observers to be correct was set on a second, separate dial. The comparison between the two settings was then recorded, as shown opposite, for each observer and for each test instance.

In all, twelve different dazzle designs, reduced down from an initial twenty-four, were tested for their effectiveness at contriving bearing distortion. They were applied to plaster models, drawing from a palette of twenty-one pigments, equal and identical to those designated by the US Navy's camouflage design team. As certain brilliant colours were discarded, only fourteen of the pigments were actually used, including black, white and grey-white.

The tests were conducted using three different, independent categories of observer. Each of their observations of each model, made at two different angles and under four different skies, was restricted to a maximum of around thirty seconds, considerably more time than a U-boat commander would have had.

The first observer was a single, highly experienced European naval lieutenant who was familiar with periscopes, ship rangefinders and weather conditions at sea. The second was a person affiliated to the test programme, who had participated closely in the preparation of the models and who, from such frequent handling, could be considered to have acquired intimate knowledge of the selected designs.

The third category comprised four separate individuals, none of whom had any foreknowledge of the dazzle designs applied to the models. They were, however, conversant with the principles of perspective and optical illusions. They were also knowledgeable about ships' structures and had experience of variations of visibility and weather at sea.

Since the trial was conducted under artificial constraints, it was determined that, for it to be deemed valid, a severe though arbitrary target of achievement should be specified. This required an apparent deviation from the correct bearing of not less than 18 degrees in order for a particular design to be judged as sufficiently potent. Previously, Norman Wilkinson had suggested that an error in course judgement of 8 degrees was sufficient to interfere with an accurate torpedo launch, while German submariners trained to incorporate an error allowance of 10 degrees in course and 2 knots in speed when plotting the position of a target ship.

In the event, the average error recorded in the study for the judgement of bearing was well in excess of these values. The variance from the true course angle of the majority of the estimates ranged on average from 20 to 30 degrees, more than enough to result in a mistaken torpedo gyro setting.

Despite amassing data that was overwhelmingly favourable, Leo Blodgett ended his thesis modestly enough with the words, 'The results of the foregoing experiments would seem to indicate that the [dazzle] method of formulating designs was an advance on preceding [low visibility] works and that the efforts were conducted in the right direction.'

Design 1.

	First Observer			Second Observer			Third Observer		
	Actual	Est.	Error	Actual Angle	Estimated Angle	Error Angle	Actual	Est.	Error
Clear	140	110	30	64	280	216	50	64	14
	60	190	130	52	68	16	122	90	32
Hazy	135	122	13	54	44	10	52	120	68
	58	110	42	128	56	72	130	240	110
Storm	112	62	50	128	72	56	135	42	93
	75	60	15	58	72	14	52	50	2
Cliffs	112	66	46	46	110	64	51	52	1
	112	66	46	126	66	60	126	136	10

Average angle of error 50 degrees.

Average of three sets of observations for each sky. Results used in plotting.

	Clear	Hazy	Storm	Cliffs.
Degrees.	38	53.5	73	38

Average angle of error assuming that with a white or black background no error would be made.   36.

An example of the observation results recorded in Leo Blodgett's study. These figures were arrived at by averaging each observer's perceived errors of bearing for each model. The sum of these averages was then averaged again to reach a final angle of deviation for each dazzle design, correlated as a measure of its effectiveness. (MIT)

Only much later, after the Millennium, were further analytical studies carried out to understand the behaviour of dazzle, having the advantage of computer synthesis to analyse rendered 3D digital images in simulated atmospheric conditions. These also produced favourable results for size and speed assessment, but for all their vindication of the disruptive dazzle concept, they took place long after the event and had little practical value in an age where radar and advanced satellite tracking systems had rendered all ship camouflage anachronistic.

Leo Blodgett went on to become an accomplished marine engineer, partner of the firm Ellis-Blodgett Associates, and, later, was the editor of the Proceedings of the Society of Naval Architects and Marine Engineers.

Between the wars, occasional supplementary consideration of the roles and applicability of different types of marine camouflage formed the basis of further official studies and reports. The US Naval Research Laboratory released two reports after some additional sea trials, 'The Preliminary Report on Low Visibility' in September 1935 and 'The Preliminary Report on Dazzle' in August 1936, but while they lent balance to the comparison of systems, they mainly repeated earlier findings. Harking back to the opinions expressed in the US Navy's 1920 reports, they noted that no single colour or low-visibility arrangement would work in all circumstances, while disruption's main value was as a defence against submarine attack where the enemy point of view was at sea level. Echoing Norman Wilkinson, the latter report added the comment that course deception patterns should neither be too extreme nor too modest in seeking to contrive perspective distortion. Also, the application of reversed perspective distortion, Frederick Waugh's innovation, giving the illusion that the stern of an approaching ship was nearer than its bow, was considered to be the most effective strategy.

It is believed that some of these deductions formed the basis of the Bureau of Construction and Repair's 'Handbook on Ship Camouflage' (Bu-C&R4), published a year later. While 'new course deception designs' were discussed within it, the only three measures advocated, complete with their paint mixing formulae, were of the low-visibility type: Dark Grey (later US Navy Measure 1), Standard Navy Grey (later the equivalent of US Navy Measure 3) and Ocean Grey (later US Navy Measure 14).

Until the late 1930s, the UK had no cause to believe that it would need to revert to camouflage to protect its naval and mercantile fleets from attack and little attention had been paid by the Royal Navy to such matters. As a result, on the outbreak of war in September 1939, the UK was as unprepared in respect of painted protection measures as it had been twenty-five years earlier and, in effect, it became necessary to begin practices and procedures virtually from scratch.

Right. Top to bottom: This scheme worn by the 'Fiji'-class cruiser HMS *Uganda* is a typical freelance paint job from early in the Second World War. Patchy in its arrangement and distribution of pattern, it would have provided only a limited degree of painted protection. It is understood that the very first British Naval vessel to adopt impromptu painted camouflage, as early as December 1939, was the destroyer HMS *Grenville*. Just over a year later, in January 1941, the first Royal Navy ship to be painted in an approved Admiralty camouflage scheme was the battleship HMS *Queen Elizabeth*. (Adrian Vicary); An unidentified French anti-submarine sloop of the *Aisne* type, thought to be *Somme*, seen between the wars in a dark, mottled form of background-blending disruptive painting. The reason for adopting camouflage at this time is not known, nor whether it was retained into the Second World War. (Alain Croce collection); Unlike the ships of the Royal Navy which, in the absence of official guidance, had largely to do their own thing until well into 1940, many official US Camouflage Measures were already prescribed in instructional directives in advance of the USA becoming involved in the Second World War. Here, USS *Wainwright* is seen on 22 March 1942 in the modified Measure 12 graded system, which had been introduced on 15 October of the previous year. (US National Archives)

7

ADAPTATION AND REGULATION – REVIVAL OF SHIP CAMOUFLAGE IN THE SECOND WORLD WAR

Despite presumed reservations about its effectiveness and a claimed lack of substantive evidence to demonstrate positive benefits from the use of dazzle or any other disruption or confusion paint scheme introduced in the First World War, many comparable practices resurfaced in the Second World War, employed by virtually all the combatant navies engaged in this new conflict. Much the same was the case in regard to the concealment practices that had been in use but abandoned prematurely in 1918. So, was this full vindication of all that had been proposed and implemented before?

The main difference was that, in general, camouflage schemes were applied to a far greater extent on warships and auxiliaries and significantly less so on merchant vessels. Another important distinction was that the treatments applied in this new conflict by the UK and USA were subject to appreciably greater regulation and control in respect of their authorisation of use, and in the designs, pattern definitions and paint pigment formulae.

The needs and objectives of camouflage at sea were no longer the same either, demanding solutions to suit a wider range of circumstances not previously encountered. During the First World War, the fighting between opposing armies had been predominantly on the battlefields of Europe with the route for foodstuffs, troops and armaments principally across the North Atlantic to French and British ports. The Second World War, by comparison, was a truly global war involving naval operations in almost every ocean and in all latitudes from the polar north, through the tropics to the southern oceans. The huge variation in the quality of the light and range of atmospheric conditions confronted required a more extensive portfolio of camouflage solutions for the ships operating in such diverse regions.

Similarly, the nature of the threat to shipping had changed. The range of submarines and the performance of their torpedoes had improved immeasurably. Naval gunnery and fire control had also been developed between the wars, but the greatest threat to emerge in the Second World War was military aviation and the danger of aerial observation or attack.

Twenty years earlier, flying machines had still been in their relative infancy, presenting only a minimal threat as an offensive weapon except, perhaps, against each other. The truly rapid speed of aircraft development since 1919, extending operational ranges and increasing weapon payloads, had been extraordinary. Naval aviation had likewise come of age and aircraft carriers permitted operations far from land, exposing ships to a new attack danger.

The question is, to what extent in these circumstances had the camouflage measures developed and adopted in the First World War been reintroduced? Inevitably, some development and refinement was to be expected, but had certain concepts been discarded entirely while others were retained as continuing to offer potential protective value? And what of the continuation of research into ship camouflage practices?

Although the USA did not enter the Second World War until December 1941, two years after it had broken out in Europe, the US Navy authorities had accelerated camouflage plans well in advance of its commitment to war, having an existing, active research station at Washington DC. From 20 June 1940, the navy establishment was reorganised, with the Bureau of Construction and Repair incorporated into a new Bureau of Ships. The Camouflage Section became part of the Bureau of Ships Research and Development Branch under the direction of Captain Henry A. Ingram, with Charles Bittinger, a First World War camoufleur at the Shipping Board, heading up camouflage development. He was joined by Everett Longley Warner as his Chief Civilian Aide, but Warner's contribution was far from cursory, for many of the new, approved design drawings were signed off by him personally. And his design philosophies were also evident in the new camouflage measures, notably the concepts based on solid geometry and reversed perspective.

Twenty-five years after he had first engaged in camouflage design, Everett Longley Warner was again involved in the promulgation of protective paint measures for the naval fleet of the USA, drawing on his earlier work. He is seen here in the early 1940s with a new generation of camoufleurs employed at the Washington DC Camouflage Section. (Roy Behrens)

Despite suggestions that, on the one hand, the USA retained a bias towards disruptive schemes while, on the other, the British favoured low visibility, the truth was that both naval authorities adopted a mix of the two, different service requirements demanding this approach. On 1 January 1941, the Bureau of Ships released a new edition of the 'Handbook on Ship Camouflage' (Ships-2), updated from the 1937 original. Drawing in part on tests conducted at sea off San Diego between June and September 1939, it introduced the first formalised US Navy Camouflage Measures 1–9. Subsequent revisions and new editions added a further sixteen (Measures 11, 12, 12 Modified, 13 and 14 on 15 October 1941; Measures 10, 15–18, 21 and 22, plus 9 and 13 Revised, on 2 June 1942; and Measures 23 and 31–33 on 14 May 1943), altogether a broad spectrum of camouflage options. Among them were both high-similarity (low visibility) and high-difference (disruption) schemes but, of interest, the transition over four years of warfare was progressively from concealment to confusion.

Measures 17 and 18 had flirted with dazzle and disruption but Measures 31, 32 and 33 were profoundly disruptive in character and were extensively applied in a variety of designs in virtually all the theatres of war and on all ship types, from small amphibious craft to the largest aircraft carriers. Their purpose was to counter attacks from sea level and from the air and they also provided background-blending protection against land backdrops in the US-led island-hopping campaign across the Pacific, which characterised the war against the Japanese.

In keeping with the realisation towards the end of the First World War, the vivid primary colours that had characterised dazzle were avoided as they were thought unnecessary, no longer required to achieve the required degree of contrast. Thus, the palette of paint colours remained confined to less-saturated, subdued tones that were more sympathetic with the maritime environment.

While that was equally true of British camouflage paint colours, the overall story in the UK was somewhat different. At the outset of the war in September 1939, there was no dedicated naval camouflage research establishment. The Admiralty Research Laboratory at Teddington, Middlesex, formed in 1921 from the Admiralty Experimental Stations, had a perfunctory involvement in naval camouflage but it wasn't until late 1941 that a research unit was created expressly for this purpose. The decision whether or not to camouflage Royal Navy ships rested with the administrative authorities at the Admiralty, subject to the approval of the commander-in-chief but, given the paucity of official camouflage guidance, some ships' companies again resorted to unofficial, customised schemes pending the definition of approved regulations.

As for merchant ships, a policy decision reached at the outset stipulated that there was no intention for them to be camouflaged disruptively or in any other patterned variant. Captain Cedric Swinton Holland conveyed this directive on behalf of the Admiralty in his Aide-Memoire No.78, 'Notes on Camouflage for Merchant Ships', dated 31 October 1939. It expressed the view that camouflage did not help to make ships less visible and that, by the time a disruptive pattern could be clearly discerned by a submarine, it would already be so close that the camouflage could

neither disguise the course nor the speed. The order to adopt Dazzle, made back in 1918, had only been rescinded in 1937, but Holland made it clear that official policy was for it not to be reinstated, not because of the criticisms he had outlined but because the expense of the paint materials would be very great, given the large size of the British merchant fleet. Thus, for British merchant ships, including troopships, plain grey overall would be the treatment and that was more or less how it remained, even with the introduction of the specific Merchant Ship Sides (MSS) and Merchant Ship Decks (MSD) scheme in April 1941, along with the recommendation for funnel and mast tops to be whitened.

It is not the intention to gloss over the camouflage activities of the Axis nations and their allies, although less is known about the research and development – a central focus of this book – behind the schemes adopted by Germany, Italy and Japan. The genesis and early advancements in ship camouflage had taken place in the UK and USA and these practices were dominated by the Americans and British in both world wars. It is inevitable, therefore, that there is a concentration on the experimentation and camouflage measures of these leading nations where it all began. That said, after September 1939, Germany appeared to have stolen a march on the British, with the official use of dazzle-like disruption clearly evident from as early as the late summer of 1940 on troop carriers intended for deployment in the planned Operation Sea Lion assault on the UK.

Initially, responsibility for German naval camouflage formed part of the brief of the Technical Bureau of the Navy Office at Hamburg (Technisches Amt der Kriegsmarinedienststelle), which was directly subordinate to the Naval High Command in Berlin. The Technical Bureau designed individual camouflage schemes for virtually every naval and auxiliary vessel. The evolution of these designs relied on methods much like those employed in the USA and UK, involving a visual critique of painted models. The prescribed designs appear to have embraced a broad range of strategies for different situations, much as was the case with the Allied navies.

A novel style of disruptive camouflage applied to many converted merchant vessels, such as commerce raiders and coastal auxiliaries, was called 'flimmestarnung'. A type of flicker-splintered dazzle, it featured a random pattern of small components in various grey, green and brown tones painted over the entire ship to break up its structural appearance and, perhaps, to induce background blending. Dazzle disruption was also widely adopted for German naval vessels from as small as Räumboote (R-boats – small minesweepers) up to the largest fleet units, the battleships *Bismarck* and *Tirpitz*. Other practices for the protective colouration of German warships included hull foreshortening, painting in and out, false bow waves and ship silhouettes.

From April 1943, responsibility for German ship camouflage was transferred to the Marinegasschutz- und Luftschutzinspektion (Marine Gas and Air Protection Inspection, MG-LSI), based at Berlin-Charlottenburg but subsequently relocated from February 1944 to the lakeside town of Plön in Schleswig-Holstein. From 26 January 1943, this organisation set about the determination of a theoretical and scientific basis for camouflage in conjunction with a study undertaken by the Amtsgruppe-Marüst at the Harnack-Haus, Berlin-Dahlem. The MG-LSI also defined the first official palette of paint colours with guidelines for their application. Prior to that, individual ship commanders had generally decided for themselves the selection and mixing of paint colours to be applied to their ships.

The situation in Italy, and Japan for that matter, was rather different. Although there had been no pre-planning of camouflage, many Italian naval vessels and some auxiliaries carried protective paint schemes throughout the sea war in the Mediterranean, notably distinctive splintered 'fish-bone', 'sawtooth' and 'sun ray' radiating angular patterns, all ostensibly controlled by the Technical Department of the Regia Marina High Command (Maristat) in Rome. The suite of Italian camouflage designs, applied to both merchant vessels and warships, showed evidence of the use of parti-colouring, hull foreshortening, fake bows and a variety of other strategies, much like those employed by other combatant navies.

It has been suggested that, in reality, the Italian ship camouflage effort of the Second World War was largely pursued as a secret surveillance undertaking, under the direction of Major Luigi Petrillo of the Naval Engineering Corps. The method followed, apparently, was to use intelligence gathering and covertly obtained photographs of British ships, along with pictorial representations of First World War schemes, and either adapt or replicate them. That is only half the story, though, for

as tasked by the Chief of the Naval Staff, from November 1940, original and unusual designs did also emanate from Petrillo's organisation, with guidelines for their implementation, the first applied experimentally to the cruisers *Fiume* and *Duca d'Aosta* early in 1941. The emphasis was on camouflage for deception rather than concealment.

Also, in the summer of 1941, the marine artist Rudolfo Claudus, an Italian of Austrian descent, was commissioned by the Regia Marina to evolve a range of unique camouflage designs, commencing with those for the cruisers *Muzio Attendolo* and *Giuseppe Garibaldi*. As far as is known, no viewing theatre was used for the evaluation of any of these schemes, their effectiveness and any need for refinement was entirely dependent on feedback from visual sightings of the ships at sea.

Although Japan, like the USA, did not enter the war until late 1941, it is fitting to consider its implementation of ship camouflage practices here. Surviving records suggest that it was nominal and sporadic. Some auxiliary vessels were evidently the recipients of extensive treatment, while the majority of Japan's naval ships remained in neutral, mono-colour livery, based on green tones. A prominent figure who, it is said, was instrumental in the advancement of Japanese camouflage designs, such as they were, was Lieutenant Commander Shizuo Fukui, a former cadet at the Naval Technical Research Institute at Yokosuka. Later, he joined the staff of the Naval Technical Office in Toyama as an inspector. Fukui is credited with producing the patterned camouflage of a number of ships while working at the Kure Navy Yard, but it is evident that certain of these schemes were cribbed from US Navy dazzle designs of the First World War.

At the outset of the Second World War, British camouflage efforts were focused on the protection of civilian and RAF ground infrastructure from aerial attack, primarily installations of strategic importance such as airfields, factories, power stations and railway marshalling yards. With this emphasis on protecting essential civilian amenities, it was placed under the Ministry of Home Security as a Civil Defence Camouflage Establishment (CDCE), headed by Captain Lancelot M. Glasson.

The Research and Experimental Department took over various buildings in Royal Leamington Spa, Warwickshire, from October 1939. Expanded in February 1941, the CDCE was then redesignated the Directorate of Camouflage under the direction of Wing Commander Thomas R. Cave-Brown-Cave. Yet, up to that time and until some months later, still no provision was made for naval camouflage requirements.

It was in this vacuum of official guidance that a British initiative emerged specifically intended for escort ships protecting Atlantic convoys. First adopted experimentally, then treated as semi-official, before eventually forming part of the Admiralty's suite of approved measures, it was the Western Approaches scheme, conceived by the renowned wildlife artist Peter Markham Scott, who was serving aboard the destroyer HMS *Broke*, on which it was first tried out.

Originally composed of three very light colours, white, pale blue (WAB) and pale green (WAG), the third colour was later dropped due to a shortage of green pigment. Widely utilised, these colours suggested a validation of the credo of Abbott Handerson Thayer, subsequently endorsed by the deliberations of the Eastman Kodak team under Loyd Jones back in 1917. In practice, however, the scheme was confined to a specific purpose on smaller naval vessels engaged on escort and patrol duties in the particular zone of operation from which its title was derived.

It was not a broad-brush concept to replace disruptive camouflage, as has been intimated by some commentators, suggesting that it indicated espousal of the low-visibility concepts of Thayer and Kerr in preference to Wilkinson's Dazzle. Such views give a false impression of an approbation of a received (but false) wisdom regarding the British attitude to naval camouflage during the Second World War. Certainly championed by certain serving naval officers, the ideas of Thayer and Kerr were by then regarded more favourably, but the reality was that the Western Approaches scheme was necessarily restricted in its application.

Rather than being exclusive, it formed part of a wide range of naval camouflage measures that included forms of disruption. Moreover, service experience revealed an operational drawback and from 1942 it was, to a large extent, superseded by the Special Home Fleet Destroyer scheme, which incorporated a hint of disruption in its design. For his troubles, though, Norman Wilkinson had been sidelined in the prevailing climate, not even invited to make any contribution to the renewed marine camouflage effort.

In June 1942, the US Navy introduced its equivalent to the Western Approaches scheme, the two-colour Measure 16 Thayer System, so-named by Bittinger and Warner. It was to continue in use throughout the Second World War, albeit limited, like its British version, to application on those ships engaged in ocean patrol and convoy escort duties in northern climes.

It was hinted, at one point, that the Cunarders *Queen Mary* and *Queen Elizabeth* should be painted in the Western Approaches scheme, but they never were – nor for that matter were any other merchant ships. Thayer's original concept had never been subjected to practical trial on cargo vessels during the First World War and one can only speculate, therefore, as to whether it would have delivered a truly effective means of protection for the exposed ships of the Atlantic convoys which, for a second time, had become the prime target. Nonetheless, the benefits of the Western Approaches scheme for its intended purpose were self-evident and its potency, as applied to patrol vessels and escorts, gained widespread recognition once it had been endorsed by the naval authorities. The Flag Officer, Northern Ireland, Sir Charles Ramsay, wasted no time in enthusiastically authorising the painting of all naval ships of the escort and patrol flotillas in accordance with Peter Scott's recommendations.

The growing realisation in the UK that there was an urgent requirement for an official service exclusively devoted to ship camouflage research eventually led to the creation of such an organisation. Lieutenant Oliver Grahame Hall RNVR, who was attached to the Training and Staff Duties Division of the Admiralty's Directorate of Scientific Research, was tasked with overseeing the assembly of a team of RNVR personnel expressly for this purpose, arising from which the Royal Naval Camouflage Section came into being around October 1941. Although it remained under the jurisdiction of the Admiralty's Naval Research Laboratory at Teddington, it formed part of the overall organisational structure of the Home Office's Directorate of Camouflage.

Consequently, the Naval Camouflage Section was also located at Leamington Spa, taking over occupancy of the former art gallery and museum in Avenue Road. Appointed as its head was Lieutenant Commander James Yunge-Bateman, known for his preference for disruptive camouflage. His team comprised three camouflage officers, five senior technical assistants, three junior technicians and a solitary model maker who had a workshop in the basement of the adjacent technical school.

The unit's strength was supplemented by specialist personnel transferred from the research laboratories of the General Electric Company (GEC) at Wembley, Middlesex, where they had been active since 1939 in studying wartime visibility problems and had commenced the development of a disappearance range gauge. Among those relocated was the talented physicist Alphonse Emile 'Bill' Schuil, who specialised in optics and the behaviour of light. Another key individual engaged in this new initiative was Robert Yorke Goodden, a liaison officer based at the Admiralty Headquarters Organisation in Bath, Somerset, who developed a range of official paint colours and was largely responsible for the Admiralty publication 'Camouflage of Ships at Sea' (CB.3098R), released in May 1943.

Reflecting its Royal Navy emphasis, the section's principal brief was to analyse the factors that contributed to making warships conspicuous while simultaneously concentrating on delineating schemes for the painted protection of every sea-going naval vessel. Despite producing countless disruptive designs, according to Admiralty records the focus was meant to be on reducing visibility range to the minimum. Almost as an afterthought, the section was also tasked with consideration of the camouflage requirements of merchant ships.

Despite its late start, the Naval Camouflage Section was well equipped, with model testing and viewing resources that were second only in their sophistication and scale to the complex facility established in the Eastman Kodak Physics Department from 1917. It is understood that, initially, a viewing range was set up over a shallow water tank constructed by the marine artist Frank Mason and equipped with a set of reversed naval binoculars to induce an impression of distance. Alphonse Schuil customised a haze filter device for this set-up.

Subsequently, that arrangement was replaced by a large, room-sized viewing theatre and broad 30ft × 21ft water tank, developed by Wilfred Shingleton, an Oscar-winning film set designer and art director (for *Great Expectations*). A new optical system, designed by Schuil, used a collimating lens (a highly corrected lens with multiple elements) to allow models to be focused on at infinity with little or no parallax (image misalignment), spherical (peripheral image curvature) or chromatic (edge fringe through refraction of colour wavelengths to different focal points) aberrations.

GENERAL ELECTRIC COMPANY NAVAL CAMOUFLAGE VIEWING RANGE

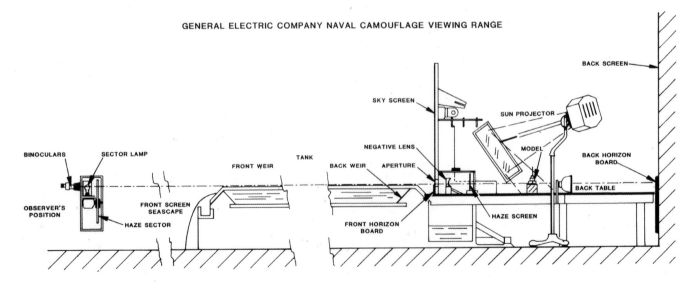

A drawing of what is thought to be the original Leamington Spa Naval Camouflage test theatre showing Frank Mason's tank, the viewing system of binoculars and the haze sector. (General Electric Company)

Besides the indoor tank in the east gallery, a second, outdoor, tank was installed between the gallery and the school to allow assessments to be carried out under natural light. Observation trials were also conducted at sea under the leadership of Oliver Grahame Hall. Sadly, during one of these operations Alphonse Schuil was killed when his vessel, the submarine P615, was torpedoed and sunk by U-123 off Freetown, Sierra Leone, on 18 April 1943. Apart from the viewing range optical system, Schuil's other major achievement was the Schuil Mark II Daylight Telephotometer.

Clockwise from left: A section of the viewing range table of the Leamington Spa test theatre. (General Electric Company); The floor plan of the Leamington Spa Naval Camouflage research facilities as subsequently developed, showing the large room-sized indoor tank and the outside, daylight viewing tank. (William Glasson, courtesy of Leamington Spa Art Gallery and Museum, redrawn by David F. Hutchings); The Leamington Spa test theatre's haze sector box. (General Electric Company)

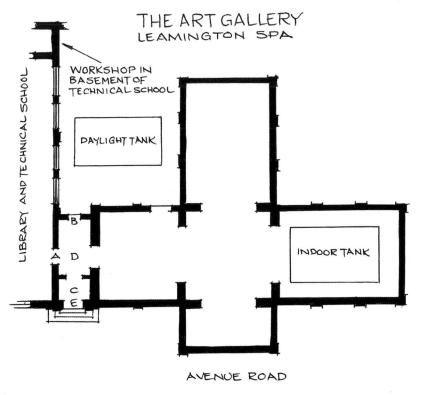

General view of the large Leamington Spa indoor testing tank, revealing its considerable expanse. Of all the model-viewing facilities created in the First and Second World Wars, only that commissioned at Eastman Kodak also utilised a water tank so that models could be evaluated floating on a realistic surface rather than on a painted seascape. The models constructed at Leamington Spa were based on enlarged plan view and elevation drawings from *Jane's Fighting Ships*. The optical system employed was designed by Alphonse Emile Schuil. The theatre was equipped with comprehensive lighting and weather simulation controls. (Kurt Hutton, Getty)

A wartime device whose conception and advancement owed itself to John M. Waldram, an illuminating engineer and colleague of Alphonse Schuil at the GEC Research Laboratories, was the Disappearance Range Gauge, shown here. Development continued after the war had ended when it was patented by GEC. (Meteorological Office)

It was more than a year after the war had ended before the Naval Camouflage Section at Leamington Spa was finally disbanded, and the museum and art gallery were vacated on 13 August 1946.

So, what conclusions can be reached from the Second World War ship camouflage practices relative to what had happened in the First World War? First and foremost, revived rather than replaced, they did draw extensively on the concepts introduced in the earlier conflict, not only their broad principles but, in many cases, their detailed design attributes – a clear vindication of the pioneering efforts of Thayer, Kerr and Mackay as well as Wilkinson and Warner. The tables opposite, of British and US surface ship schemes from the Second World War, indicate the links between the 'new' practices and those adopted or proposed from 1917 onwards, the former being virtually identical or close lookalikes to their predecessors. Many of these Admiralty and US Navy schemes incorporated counter-shading as had been recommended by Thayer, Brush, Toch and Herzog. Equally, the tables reveal the wide diversity of authorised camouflage options, demonstrating unequivocally that there was no universal panacea, no single one-size-fits-all ship camouflage scheme that suited all circumstances. It was definitely not a matter of low visibility versus dazzle or disruption – or versus any other concept, for that matter. It was a case of all stratagems having a value, whether for a particular threat or a particular warzone.

Reflecting back on the First World War experience, similar outcomes could perhaps have been reached then, but for the determination to

Admiralty Surface Ship Camouflage Schemes, Second World War	
SCHEME TITLE	FIRST WORLD WAR EQUIVALENT
Western Approaches Scheme [1940] devised by Sir Peter Scott	Thayer Low Visibility
Mountbatten Pink [1940] devised by Lord Louis Mountbatten	See Note below
Standard Grey Scheme - Merchant Ships, MSS & MSD [1941]	
Light Disruptive Patterns [1942]	Dazzle Disruption
Intermediate Disruptive Patterns [1942]	
Dark Disruptive Patterns [1942]	
Special Home Fleet Destroyer Scheme [1942]	Low Visibility Disruption
Alternative Schemes [1942]	
Standard (Light Tone) Schemes [1944]	Fake ship shapes (basic)
NOTE: In April-May 1917, HMS *Ramillies* was experimentally painted a shade of pink as described by H.M. Le Fleming and as portrayed in Charles Pears' painting but it was a shortlived experiment, replaced by dazzle in 1918.	

A list of the Second World War naval camouflage schemes for the surface ships of the Royal Navy. (Author)

A list of the Second World War naval camouflage schemes for the surface ships of the US Navy. Whereas a range of low-visibility schemes was deployed during the Second World War for use in particular zones of operation, the vast majority of naval vessels of most nationalities were painted disruptively in one form or another. In total there were well over 100 US Navy disruptive designs, further modified for each Measure and each ship type. (Author)

United States Navy Surface Ship Camouflage Measures, Second World War		
MEASURE No	MEASURE TITLE	FIRST WORLD WAR EQUIVALENT
1	Dark Grey System [1941]	
2	Graded System [1941]	
3	Light Grey System [1941]	
4	Black System for Destroyers [1941]	
5	Deception System: false bow waves [1941]	False bow waves
6-8	Deception Systems: painted disguise to simulate other ship types [1941]	
11	Sea Blue System [1941]	
12	Graded System - Basic & Modified (mottled) [1941]	Kerr Parti-colouring
13	Haze Grey System [1941]	
14	Ocean Grey System [1941]	
15	Disruptive System (experimental) [1942]	Dazzle Disruption
16	Thayer System [1942]	Thayer Low Visibility
17	Dazzle System (prototype for Measures 31-33) [1942]	Dazzle
18	Graded System (comparable to Measure 12 with Haze Grey replacing Ocean Grey) [1942]	Kerr Parti-colouring
21	Navy Blue System [1942]	
22	Graded System (with false horizon simulation) [1942]	
23	Light Grey System [1943]	
31	Dark Pattern System [1943]	Disruption (USN) & Background Blending
32	Medium Pattern System [1943]	
33	Light Pattern System [1943]	

prove one concept over another. Views to that effect, expressed at that time and from the Second World War period, most pertinently amplify this point. In a supplement to the 'Bulletin of Submarine Warfare' from 17 May 1918, a French expert had this to say:

It is incorrect to claim that an object on the sea painted in such or such a way is any more or less visible. Nothing is positive on this question and an object if it is rendered less visible for a definite reason in a certain light will be rendered more visible for the same reason in another light.

This reasoning applied equally to disruptive colouration, since there was always going to be a trade-off between increased visibility and interference with perception, especially in certain light conditions or as seen from less favourable angles of view. The Admiralty concurred with this in the CB.3098R document of May 1943: 'Ships could not just manoeuvre into positions that best suited the effects of the camouflage they were painted in.'

Everett Warner expressed his agreement with these opinions in his own apposite contribution: 'I do not believe in fixed systems of camouflage. It should remain as flexible a strategy, always ready to be altered to meet either permanent or temporary changes in conditions.'

If there was one camouflage variant that predominated, it was the core concept of the hybrid approach that had been given only cursory consideration in the First World War. Many of the patterned schemes

VIII. REPORTS ON OBSERVATIONS OF CAMOUFLAGED SHIPS

26. Reports on observations of camouflaged ships are required both from the air and from the surface. As knowledge of the exact conditions of light and background is essential in assessing the reports, reports should be made in the following form:--

Form of Reports on Camouflage Ships
A. Name or description of target ship.
B. Approximate position.
C. Time and date of observation.
D. Weather:--

    (i) Proportion of cloud.

    (ii) Whether sun or moon was unobscured, partially obscured or obscured by cloud, fog or haze.

    (iii) Approximate angle of elevation of sun or moon, if any.

E. Height from which target ship was observed.
F. Whether observed against sky, sea or land background, stating tone and colour of background and state of sea.
G. Approximate compass bearings of target ship and sun or moon from observer, range of observation and course of target ship, shown diagrammatically of C.A.F.O. diagram 76/42:

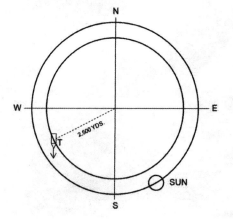

Just two years prior to painted camouflage becoming effectively superfluous through the increasing use and greater sophistication of radar systems, the Admiralty for the first time introduced a system of observation reporting. Announced in the publication 'Camouflage of Ships at Sea' (CB3098R), it used the diagram 76/42, first published in 'Confidential Admiralty Fleet Order' (CAFO) 1112. Intended to assist the improvement of the camouflage measures then in force, it is not certain what benefit derived from this action at such a relatively late stage in the hostilities, besides which the first of the 'Alternative Schemes to the Disruptive Patterns' had been added to the naval camouflage arsenal even as the report system was being implemented. (Admiralty)

employed between 1941 and 1945 sought, in their colour and pattern mix, to achieve a combination of reduced visibility along with a measure of course misperception, depending on the distance at which they were observed.

The principal objectives of ship camouflage also remained unchanged – low visibility, disruptive course distortion, background blending and structural flattening. There were improvements and new designs, but essentially all derived from the established stratagems. Nothing in the form of truly new, original concepts emerged, although briefly in the UK an idea was flirted with to create a chameleon-like system using so-called reflective/refractive coatings, but this was discarded at the embryonic stage.

As for the 'table-top' evaluation of camouflage systems and designs, this certainly continued in the Second World War in the USA, UK and Germany, although in the latter case it may have had as much to do with seeking theoretical rather than practical outcomes.

A question of immense significance that remains unanswered is what benefits may have accrued had British and Commonwealth merchantmen been painted in a form of patterned camouflage as they had been from 1917 onwards? In the Battle of the Atlantic, the campaign which most closely resembles the submarine war of 1914–18, a total of 2,603 merchant ships were lost between 1939 and 1945 at a cost of 30,248 seamen's lives. Although the losses of ships were less than half those sustained in the First World War, the human casualties were more than double the 14,661 who had been killed in the earlier period. In 1939, the Admiralty had rejected the use of patterned camouflage for merchant ships on the grounds of cost, placing its reliance instead on the convoy system, even though German wolfpack tactics were to prove capable of nullifying its advantages.

The Western Approaches scheme was extolled as the most effective camouflage system for northern waters, but if it had worked so unquestionably well on naval ships, why was it not also applied to the unarmed merchant vessels of Atlantic convoys? It would not be unreasonable to conclude that, had such an additional form of protection been employed, the severe loss of life that occurred may well have been appreciably reduced. Or was the cost of paint more important than the cost of human lives?

The illustrations that follow have been selected to demonstrate the close similarities of the ship camouflage schemes and measures of the Second World War to those that originated in the First World War. They reveal that, rather than the earlier practices having been superseded, most were in fact revived as virtually identical or as closely similar.

Thayer-type low-visibility disruption:

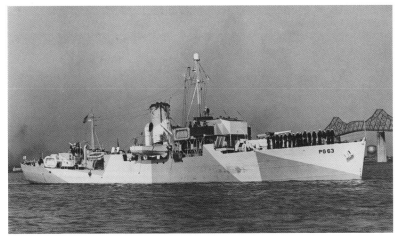

Clockwise from top left: The Lend-Lease destroyer HMS *Rockingham*, formerly USS *Swasey*, in the original three-tone Western Approaches scheme. Typical of many black and white photographs, the contrast between the colours appears more pronounced than it actually was. The mixing specifications show that WAB (pale blue) and WAG (pale green) had almost identical reflection factors. (Crown Copyright); An example of the Measure 16 Thayer System of low-visibility camouflage adopted by the US Navy in 1942 and here applied to the US Coast Guard vessel *Spencer* engaged in convoy escort duties. (US Coast Guard); The flotilla leader HMS *Keppel* seen in the Western Approaches two-colour variant, an image that reveals how light the scheme was. Devised by Sir Peter Scott, the Western Approaches scheme drew closely on Abbott Handerson Thayer's concepts, which were also manifest in the contemporary US Navy's Measure 16. (Crown Copyright); The patrol corvette USS *Surprise* seen in Measure 16, the Thayer system of white and blue, which represented a vindication of Abbott Handerson Thayer's radical concept. (US National Archives, 19-N-66636)

Kerr's Parti-Colouring and Mackay's Small Pattern with wave simulation:

An unidentified Carlisle-class cruiser sports a two-tone splotch design quite possibly influenced by the Parti-Colouring scheme advocated by Sir John Graham Kerr, although it was not necessarily intended to interfere with the rangefinding accuracy of enemy surface vessels. (Imperial War Museum, A4080)

The destroyer USS *Meade*, photographed on 20 June 1942, is in the modified US Navy two-tone Measure 12. She appears to represent a combination of the Kerr Parti-Colouring and Mackay Low-Visibility concepts. (US National Archives, 19-N-30842)

USS *Hobson* shows off another example of the modified Measure 12 low-visibility splotch system, which clearly replicates Kerr's Parti-Colouring scheme. (US Naval History and Heritage Command, NH53548)

Prior to the introduction of the US Navy's Patterned Measures, the Measure 12 modified system was widely applied to naval vessels of all sizes and types, including battleships, aircraft carriers and cruisers, as shown here on USS *Cincinatti*. (US National Archives, 19-N-32590)

Seen after Japan's surrender, the heavy cruiser HIJNS *Myoko* reveals that certain ships of the Imperial Japanese Navy were also painted in splotch camouflage. (Shizuo Fukui)

The former Rotterdam Lloyd passenger liner *Baloeran* was renamed *Strassburg* after she fell into German hands. Her camouflage design also hints at Kerr's Parti-Colouring concept. (Author's collection)

The amphibious vessel USS *Endymion* carries a speckled camouflage pattern with vaguely undulating contours, to some extent reminiscent of William Andrew Mackay's pixelated Low-Visibility scheme of 1917. In fact, it is Design 18L of disruptive Measure 31. (US National Archives)

Disruption and dazzle:

Operational difficulties experienced with the Western Approaches scheme led to the introduction of the Admiralty Special Home Fleet Destroyer scheme, a more disruptive derivative. It is seen here applied to the Canadian Tribal-class destroyer HMCS *Huron*. Its purpose was not so much to confuse the enemy as to aid station-keeping. (Canadian Department of National Defence)

The Luzon-class repair ship USS *Oahu* has a camouflage pattern which gives the impression of a typical Warner First World War disruptive design. The colours of this Measure 31, Design 6Ax scheme are Ocean Green, Navy Green and black. (US National Archives, 19-N-66540)

A more typical dazzle type of disruptive camouflage was evident on the ships of many of the combatant nations in the Second World War, as here on HMS *Hawkins*. (Crown Copyright)

Italian battleship RIS *Roma*. (ANSA, Marina Militare)

Italian light cruiser RIS *Scipione Africano*. (Ufficio Storico)

The former Norddeutscher Lloyd ocean liners *Bremen* and *Europa* were converted to troopships in readiness for the planned Operation Sea Lion (*Seelöwe*) – the invasion and occupation of Britain. Both ships were painted in disruptive camouflage sometime between Hitler's Directive No. 6 of 16 July 1940, in which he signified his intentions, and 16 September 1940, the date on which the assault was scheduled to be launched. The photograph shows *Bremen* during rehearsals for the operation. (WZ-Bilddienst)

USS *Haynsworth*, painted in Measure 31 dark disruptive scheme in the Design 16D variant, clearly demonstrates the retention of obtrusive disruption camouflage in the Second World War. The design is reminiscent of Waugh's scheme, painted in 1918 on the American cargo ship *West Mahomet*. (US Naval History and Heritage Command, NH45510)

This unidentified Italian transport is painted in a camouflage design much like many of the intensively striped British Dazzle schemes of later in the First World War. (Archiva Panstowe)

Germany seems to have been one of the few countries that attempted to develop a new form of painted camouflage. Effectively, though, the 'flimmestarnung' splinter camouflage and the schemes on the two captured French ships shown in this and the following photograph were derivative, all having their roots in First World War dazzle. This is the French Line's *De La Salle* captured in 1940 and redesignated as *H17*. (Alain Croce collection)

Elsas (formerly *Cote d'Azur*). It should be remembered that no camouflage scheme of either the low-visibility variety or disruption or dazzle worked in practice if the ship was viewed between the observer and the source of light. Ships are then seen as silhouettes, regardless of how they are painted. Conversely, in dull, misty or overcast weather, all ships at a distance, whether painted for low visibility or disruption, benefit to some degree from a measure of natural concealment, regardless of their paint scheme. (Alain Croce collection)

3D solid geometric dazzle:

The return of Everett Longley Warner to the US Navy's Camouflage Section in the Second World War provided the opportunity to revive his ideas on the use of solid geometry for disruptive designs, a number of which were introduced. This is the cruiser USS *Trenton* in Measure 33, Design 2F, photographed in the Gulf of Panama on 14 July 1944. (US National Archives, 19-N-91697)

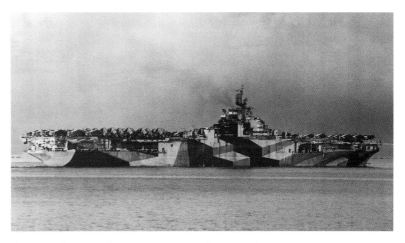

The Essex-class aircraft carrier USS *Hancock* is painted in Measure 32, Design 3A, which closely resembles the First World War Type 9, Design K camouflage. (US National Archives, 80-G-294131)

Another American fleet carrier, USS *Ranger*, has Measure 33, Design 1A camouflage, another solid geometric composition. (US Naval Aviation Museum)

A further example of a Second World War solid geometric disruptive design is painted on the naval auxiliary USS *Zaniah*. It is Measure 33, Design 9D. (US Naval History and Heritage Command, NH84627)

Although Japan produced some unique disruptive camouflage designs, it was not averse to blatantly copying the schemes of its adversaries. Here, the auxiliary cruiser *Aikoku Maru* has an exact replica of the Type 9, Design K scheme from the First World War painted on her starboard hull. Likewise, her portside design duplicated that of the Type 9, Design K port projection. Photographs show that her sister ship *Hokoku Maru* carried the same designs, but on her, they were reversed port to starboard. (Shizuo Fukui)

Reversed perspective:

A particularly good example of the revived reverse perspective distortion is seen on aircraft carrier USS *Franklin* in Measure 32, Design 6A. This design employs quadrant shapes, progressively scaled down in size towards her bow. When viewed at sea level, this would suggest that the stern end of the ship is nearest, giving the impression she is bearing away to starboard. (US National Archives, 80-G-224597)

In this Measure 33, Design 16D scheme, the heavy cruiser USS *Baltimore* shows off another approach to conveying reversed perspective along with a pronounced foreshortening device on the midships section of her hull. These measures reflected a continuing concern with the undersea menace, especially in the Atlantic Ocean. (US Naval History and Heritage Command, NH91462)

Hints of 'Fleet System':

The destroyer USS *Lang* has splintered triangular shapes to disturb the continuity of her weather deck as first exhibited in the 'Fleet System' trials of 1917–18. (US Naval History and Heritage Command, NH91383)

Another example of this deck edges interference device from the 'Fleet System' trials can be seen in a more ragged style along the hull of the cruiser USS *Nashville*, photographed on 1 April 1942. In both cases, these were actually derivatives of the modified Measure 12. (US National Archives, 19-N-28993)

Background blending:

The amphibious warfare, high-speed transport USS *Crosley* reveals the background-blending variant of Measure 31, Design 20L. The colours are greens and browns. (US Naval History and Heritage Command, NH91585)

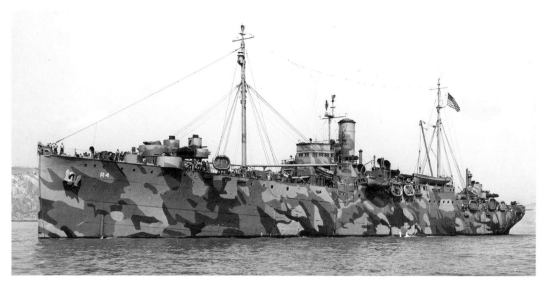

Also painted in Measure 31, Design 20L is the repair ship USS *Vestal*, photographed on 8 September 1944. (US National Archives, 19-N-71579)

The German '*flimmestarnung*' concept of splinter dazzle in three or four shades of grey was a new take on disruptive painting in the Second World War. As coastal mine vessels, like the *minenschiff* shown here in 1942, were its main recipients, its purpose was primarily background blending. (Edward Wilson collection)

Ship camouflage as practised in the First and Second World Wars has been described as the greatest ever display of public artwork. To many, when it first appeared its purposes were a mystery, and it remains a topic of fascination. It is to be hoped that in these pages it has been afforded appropriate explanation, while its manifold idiosyncrasies have, at the same time, been revealed in all their variety.

BIBLIOGRAPHY

BOOKS AND ARTICLES

Bates, Lindell T., 'The Science of Low Visibility and Deception as an Aid to the Defense of Vessels Against Attack by Submarines' (Submarine Defense Association, 8 March 1918).

Bates, Lindell T., 'Ship Camouflage Behind the Scenes', camoupedia.blogspot.com.

Behrens, Professor Roy, 'Optical Science Meets Visual Art: The Camouflage Experiments of William Andrew Mackay', camoupedia.blogspot.com.

Behrens, Professor Roy, *Ship Shape – A Dazzle Camouflage Source Book* (Bobolink Books, 2012) (contains the articles below, marked ★):

★Bement, Alon, 'Principles Underlying Ship Camouflage', *International Marine Engineering* (February 1919).

★Jones, Loyd Ancile, 'Theatre for Studying Camouflage Ship Models', *Transactions of the Illuminating Engineering Society*, Vol. 14 (21 July 1919).

★Skerrett, Robert G., 'Hiding Ships with Paint', *Popular Science Monthly*, Vol. 92 (1918).

★Thayer, Abbott Handerson, 'Disruptive Camouflage', *Scientific Monthly*, Vol. 7 (1918).

★Thayer, Abbott Handerson, 'Teaching Britain His Job – The Best Ship Color is White', *New York Tribune* (13 August 1916).

★Warner, Everett Longley, 'Painting Battleships for Low Visibility', *Transactions of the Illuminating Engineering Society*, Vol. 14 (21 July 1919).

★Warner, Everett Longley, 'The Science of Marine Camouflage Design', *Transactions of the Illuminating Engineering Society*, Vol. 14 (21 July 1919).

★Yates, Raymond Francis, 'The Science of Camouflage Explained', *Everyday Engineering Magazine* (March 1919).

Blodgett, Leo S., 'Ship Camouflage' (Naval Architecture and Marine Engineering Department thesis, Massachusetts Institute of Technology, 12 May 1919).

Forbes, Peter, *Dazzled and Deceived: Mimicry and Camouflage* (Yale University Press, 2009).

Gardiner, Robert, and David Brown (eds), *Conway's History of the Ship: Eclipse of the Big Gun*, Chapter 14: 'Camouflage and Deception' (Conway Maritime Press, 1992).

Goodden, Henrietta, *Camouflage and Art: Design for Deception in World War 2* (Unicorn Press, 2007).

Gordon, Jan, 'The Art of Dazzle Painting', *Land & Water* (12 December 1918).

Jones, Loyd Ancile, *Protective Coloration as a Means of Defense Against Submarines* (Submarine Defense Association, 8 March 1918).

Kerr, Professor John Graham, 'Camouflage of Ships and the Underlying Scientific Principles', *Transactions of the Thirty-Fifth Session of the North East Coast Institute of Engineers and Shipbuilders* (10 July 1919).

Leamington Spa Art Gallery and Museum, *Concealment and Deception: The Art of the Camoufleurs of Leamington Spa 1939–1945* (Warwick District Council, 2016).

Murphy, Hugh, and Martin Bellamy, 'The Dazzling Zoologist – John Graham Kerr and the Early Development of Ship Camouflage', *The Northern Mariner*, XIX, No. 2 (April 2009).

Saffroy, Frédéric, 'Formes, couleurs et optique: le camouflage, art ou science – L'itinéraire de Pierre Gatier' (Presse universitaires du Midi).

Saibène, Marc et al., *La Marine Marchande Française 1914–1918*, *Informer* 2011 (excerpts translated by Ralph Currell, 31 January 2014).

Sekuler, Robert, and Randolph Blake, *Perception* (3rd Edition, McGraw-Hill, 1994).

Stevens, W.R., and J.M. Waldram, 'Laboratory Techniques in Solving Wartime Visibility Problems', Communication No. 357, *Proceedings of the Illuminating Engineering Society* (General Electric Co., 1946).

Van Buskirk, Charles Harold, *The Development of Marine Camouflage and Tests Relating Thereto* (US Navy Bureau of Construction and Repair, 1 May 1919).

Warner, Everett L., 'Fooling the Iron Fish: The Inside Story of Marine Camouflage', *Everybody's Magazine* (November 1919).

Warner, Everett L., 'Ship Camouflage Manual for Pattern Design Application' (US Navy Bureau of Ships, June 1944 – unpublished).

Wilkinson, Norman, 'The Dazzle Painting of Ships', *Transactions of the Thirty-Fifth Session of the North East Coast Institute of Engineers and Shipbuilders*, 10 July 1919, and *The Nautical Gazette*, 13 September 1919.

Wilkinson, Norman, *General Directions for Dazzle Painting* (Controller General of Merchant Shipping, no date).

Williams, David Lloyd, *Liners in Battledress* (Conway Maritime Press, 1989).

Williams, David Lloyd, *Naval Camouflage 1914–1945: A Complete Visual Reference* (Chatham Publishing, 2001).

OFFICIAL REPORTS AND DOCUMENTS, CAMOUFLAGE ORDERS AND PATENTS

Admiralty (RN), 'Camouflage of Sea-Going Ships', CAFO 1112 (1942).

Admiralty (RN), 'Camouflage of Ships at Sea', CB 3098R (1943 and 1945).

Glasgow University Archives, Papers of Sir John Graham Kerr, 1869–1957: War Camouflage Correspondence (DC 006/246-365) and War Camouflage (DC 006/366-780).

The National Archives, Kew, 'Dazzle Painting of Merchant Ships', *Admiralty Committee Report*, 31 July 1918 (ADM 1/8533/215).

The National Archives, Kew, 'Quantitative Assessment of Painted Camouflage' (ADM 212/128).

The National Archives, Kew, 'Creation of Disappearance Range Gauge', 23 May 1945 (ADM 212/132).

The National Archives, Kew, 'Notes on the Effect of Counter-Shading Camouflage' (ADM 212/135).

The National Archives, Kew, 'Committee of Enquiry on Dazzle Painting of Ships: Consideration of Various Patent Claims' (ADM 245/4); also, 'Third Report', HMSO, 21 October 1924.

United States National Archives, 'Correspondence, Reports and Other Records of the Camouflage Section of the Maintenance Division Concerning Use of Camouflage in World War 1, 1917–19', Records of the Bureau of Construction and Repair, 1794–1941 (19.3.2).

United States Naval Research Laboratory, 'Preliminary Report on Dazzle', August 1936 (No H-1302).

United States Naval Research Laboratory, 'Preliminary Report on Low Visibility', September 1935 (No H-1196).

United States Naval Research Laboratory, 'The Problem of Visibility', March 1934 (No H-1036).

United States Navy Department, 'Handbook on Ship Camouflage', Bu-C&R-4, Bureau of Construction and Repair, 15 September 1937. Revised and reissued by the Bureau of Ships in 1941 as 'Ship Camouflage Instructions' (Ships-2) with further new editions, revisions and supplements through to March 1945.

United States Navy Department, 'Marine Camouflage', Bureau of Construction and Repair, file 14258-A14 (May 1920).

United States Patent Office, 'Colorimeter', Loyd Ancile Jones, Patent No. 1,496,374 (3 June 1924).

United States Patent Office, 'Means and Method for Measuring Visibility', Loyd Ancile Jones, Patent No. 1,437,809 (5 December 1922).

United States Patent Office, 'Process of Rendering Objects Less Visible Against Backgrounds', William Andrew Mackay, Patent No. 1,305,296 (3 June 1919).

United States Patent Office, 'Process of Treating the Outsides of Ships Etc., For Making Them Less Visible', Abbott Handerson Thayer and Gerome Brush, Patent No. 715,013 (2 October 1902).

Also see www.usndazzle.com.

ACKNOWLEDGEMENTS

The internet has made an enormous difference to research into subjects as complex as the development of maritime camouflage through two world wars, permitting access to historical records worldwide. However, much essential research remains dependent on personal access to original documents held by various archive institutions as well as on the ready support and contributions of correspondents, both specialist and enthusiast, at home and abroad, to all of whom I gratefully acknowledge the assistance received.

My thanks to Nereo Castelli and Marco Ghiglino in Italy, Jean-Yves Brouard in France, Marek Twardowski in Poland, and Camilla Wilkinson, Richard de Kerbrech, Nick Hawkins, David Hutchings, Kim Tomlinson and Adrian Vicary in the UK. Besides those individuals, I must acknowledge the assistance received from many associations, libraries and technical institutions, among them the Australian National Maritime Museum (Inger Sheil), Deutsches Marine Museum (Nina Nustede), George Eastman Library, Glasgow University Archives, Massachusetts Institute of Technology, The National Archives (Kew, London), UK Meteorological Office Library, US National Archives and Records Administration, US Naval History and Heritage Command, and the US Patent Office. Special mention must be made to Lily Crowther at the Leamington Spa Art Gallery and Museum and to Angelina Callahan at the US Naval Research Laboratory.

If there is one person who I would single out for my particular gratitude, for his unstinting and constant flow of ideas and contributions, it is Roy Behrens, Emeritus Professor of Art at the University of Northern Iowa. He has probably done more than anyone to promulgate knowledge about dazzle camouflage in the regular blogs on his Camoupedia website. His excellent book *Ship Shape* is the most comprehensive anthology of the camouflage work carried out in the First World War, a compendium of the writings, in their own words, of those camoufleurs who were personally engaged in the development of ship camouflage practices during that conflict. Over more than two years, Roy has provided me with original reports (some unpublished), countless articles and many images from his extensive personal camouflage library. He has willingly clarified points that I have raised in our correspondence and he has directed me to resources I would not otherwise have known about. No author could have received more generous support.

There were certain ship camouflage schemes which did not appear to fit in with the broad principles of established low visibility or disruptive thinking. Probably experimental in nature, they were so extreme in character or completely bizarre that their purpose is inexplicable.

The French cruiser *Gloire*, displaying one of the most intensely obtrusive disruptive liveries, was painted in this extraordinary fashion while she operated with US Naval forces. It creates an illusion that her hull is buckled or corrugated. She is seen off New York on 16 November 1943. (US National Archives, 80-G-K-1209)

Of the camouflage measures that defy explanation, those shown here make a particular case in point. Depicted are (left) the former Italia Line passenger ship *Conte di Savoia* and (right) her consort, *Rex*, with strange ripple markings along the lengths of their hulls. Photographed in both cases in 1940, the purpose of these unusual markings is not known. (Mario Cicogna)